NUMBERLAND

MITCHELL SYMONS

NUMBERLAND

the WORLD
in
NUMBERS

First published in Great Britain in 2013 by
Michael O'Mara Books Limited
9 Lion Yard
Tremadoc Road
London SW4 7NQ

A CIP catalogue record for this book is available from the British Library.

Papers used by Michael O'Mara Books Limited are natural, recyclable
products made from wood grown in sustainable forests. The
manufacturing processes conform to the environmental regulations
of the country of origin.

ISBN: 978-1-78243-060-5 in hardback print format
ISBN: 978-1-78243-194-7 in paperback format
ISBN: 978-1-78243-159-6 in e-book format

1 2 3 4 5 6 7 8 9 10

Designed and typeset by Ana Bjezancevic
The publishers would like to thank Oliver Redding for selected images and
www.shutterstock.com

Printed and bound by CPI Group (UK) Ltd, Croydon, CR0 4YY

www.mombooks.com

The author and publishers have endeavoured to verify all the facts and
statistics in this book with bona fide sources. Neither the author nor the
publisher can guarantee the accuracy or usability of any information
contained herein, nor accept any liability for any injury or loss that may
occur as a result of information given in this book.

CONTENTS

INTRODUCTION

Pythagoras thought that numbers rule the universe – and he was right – they're everywhere: in every tiny thing we do, from the number of bed bugs in our beds (an amazing number, go on, look it up) to the length, in metres, of all the eyelashes shed by a human in their lifetime. We don't often *think* about these numbers, but when we do, they can be incredible.

Numbers have always fascinated me and when I started collecting number-related trivia, I found that I just couldn't stop. There's something about nuggets of information that is so satisfying, and attaching numbers can bring those facts into sharper focus. For instance, who knew that every time you lick a stamp, you're consuming one-tenth of a calorie, that the human heart pumps 55 million gallons of blood by the age of seventy, that just one litre of oil can ruin one million litres of fresh water, or that there are 60,000 miles of blood vessels in the human body.

In this book I've tried to look at our everyday world from a different angle. If we stopped every now and again to consider some of the workaday things we do, see, or are a part of, we would in some cases probably be quite shocked, and it's these types of numbers I've included here. Ranging from the natural world to music, the arts, the human body, creepy-crawlies, love and marriage and beyond, *Numberland* is a book for those who are fascinated by details and the minutiae of life.

I hope you become as addicted to it as I have ...

ACKNOWLEDGEMENTS

I'd like to thank my editor Louise Dixon and her talented team, especially assistant George Maudsley and designer Ana Bjezancevic.

In addition, I'd also like to thank the following people for their help, contributions and/or support: Gilly Adams, Jonathan Fingerhut, Jenny Garrison, Philip Garrison, Bryn Musson, Nicholas Ridge, Charlie Symons, Jack Symons, Louise Symons, David Thomas, Martin Townsend, Clair Woodward and Rob Woolley.

AROUND THE WORLD

0

The number of airports in Monaco (the nearest is Nice in France).

0

The number of continents that are wider in the south than they are in the north.

1

The number of countries crossed by both the Equator and the Tropic of Capricorn (Brazil).

1

The number of countries that officially measure their nation's total Happiness (Bhutan).

1

The number of countries in the Middle East that don't have deserts (Lebanon).

1

The number of countries that *don't* have a rectangular flag (Nepal).

1

Only one of the Seven Wonders of the World still survives: the Great Pyramid of Giza.

1

The number of countries that have every type of climate (New Zealand).

1

The number of countries that have a national flag with a Bible in it (The Dominican Republic).

1

The number of towns in the world whose names begin with ABC (the Dutch town of Abcoude).

1

The number of countries that are entirely higher than 1,000 metres above mean sea level (Lesotho).

1

The number of countries that start with a Q (Qatar).

1

The number of countries that end with a Q (Iraq).

1

The number of English-speaking countries in South America (Guyana).

1

The number of countries where you can see the sun rise in the Pacific and set in the Atlantic (Panama).

1

The number of countries with a jaguar preserve (Belize).

1

The number of underwater restaurants in the world (in the Maldives).

1

The number of countries that don't carry their name on their stamps. It is the United Kingdom because the British invented the modern postage stamp (in 1840) and so are allowed this special dispensation – but only on the understanding that the monarch's head appears on each stamp.

1.75
The smallest island with country status is Pitcairn, which is just 1.75 square miles.

2
The number of doubly landlocked countries in the world (i.e. both entirely surrounded by landlocked countries) (Liechtenstein and Uzbekistan).

2
The number of continents that Istanbul is in (Europe and Asia).

2
The number of countries that have borders on three oceans (the US & Canada).

2
The number of minor earthquakes that occur every minute somewhere in the world.

2
The number of countries with flags that have guns or rifles on them (Mozambique and Guatemala).

2
There are thousands of languages spoken in the world but it is estimated (by the UN) that a language disappears every two weeks.

2.7
After opposition calls for a boycott, voter turnout in the 1983 Jamaican general elections was just 2.7%.

3

The percentage of Germans who speak French fluently (nearly half of all Germans are fluent in English).

3

The Pacific Ocean, the largest ocean, is three times bigger than Asia, the largest continent.

3

The number of world capitals that begin with the letter O in English: Ottawa, Canada; Oslo, Norway; and Ouagadougou, Burkina Faso.

3

The number of people per square kilometre in Canada. Germany has 235 per square kilometre.

4

You could drive a car round the world four times with the amount of fuel in a jumbo jet.

4

The number of countries with one syllable in their English name: Chad, France, Greece and Spain.

4

Almost all (just under 99 per cent) Egyptians live on about four per cent of the land – near the Nile River and its delta.

4.2

The depth (in kilometres) of the deepest mine in the world – Western Deep Levels near Charletonville, South Africa.

4.3
The number of births in the world every second.

4.42
The length in miles of the world's longest canal tunnel (the Rove tunnel in the Canal de Marseille au Rhone).

5
The percentage of world wealth owned by fifty per cent .of world's population.

5
There's a Starbucks in South Korea that has five levels.

5
The number of ways the letter 'F' can be pronounced in Icelandic.

5
The number of countries in Europe that touch only one other: Portugal, Denmark, San Marino, Vatican City and Monaco.

5
The number of days it takes the Otis Elevator company to transport the equivalent of the world's population

5.54
The number of centimetres of coastline per person in the world.

6
The number of hours of sunshine the Falkland Islands gets daily during the summer (they only get two–three hours of direct sunlight per day in winter).

6
The ratio of people to TVs in the world is six to one.

7
The average North Korean seven-year-old is seven centimetres shorter than the average South Korean seven-year-old.

7
The number of different surnames on the island of Tristan da Cunha.

8.2
The length in metres of the world's longest stalactite (as found in Jeita Grotto in Lebanon).

9
The number of words in the language of the Solomon Islands for the different stages of coconut maturation.

10
The percentage of the workforce in Egypt that is under 12 years of age.

10
The percentage of Greece's islands that are populated.

10

The number of countries that Brazil borders: that's every country in South America except Chile and Ecuador.

10

The number of miles Niagara Falls has moved upstream in the last 10,000 years.

10

The percentage of Gabon's land mass that its national parks take up.

10

The percentage of people who live on islands.

10

The world's deepest post box is located ten metres beneath the waters of Susami Bay, Japan. It is used by divers and emptied daily by the post office.

11

The number of official languages in South Africa.

12

The percentage of the inhabitants of Luxembourg who are of Portuguese descent.

12

The percentage of the Earth's land surface that is permanently covered by ice and snow.

13

Ethiopia is famous as the country of '13 months of sunshine'. The Ethiopian year is based on the Julian calendar, which has 12 months of 30 days each and a 13th month called Pagume, which has five days (or six in a leap year).

13

The number of countries that have crescent moons on their flags.

14

The number of islands Sweden's capital city Stockholm is built on.

15

The number of countries that border China.

15

In Switzerland, when a man reaches 20 years of age, he is required to undergo 15 weeks of military training.

17.75

If the world's total land area were divided equally among the world's people, each person would get about 17.75 acres.

18

The number of islands in the Faroes.

20

The average number of hours per week that the people of Argentina listen to the radio.

20

Hong Kong has the world's highest ratio of restaurants: one for every 20 people.

20

The percentage of the Earth that is permanently frozen.

20

The percentage of road accidents in Sweden that involve a moose.

20

The number of minutes it takes to make the world's shortest inter-continental flight (from Gibraltar to Tangiers, Northern Africa).

20

The number of countries in the world where under half the population is literate.

20

The Angel Falls in Venezuela are nearly 20 times taller than Niagara Falls.

21

Brunei's lowest-recorded temperature in degrees Celsius.

23

Japan has one vending machine for every 23 citizens – selling everything from drinks to food to iPods.

24

The average depth of the Earth's crust in kilometres.

25

The percentage of food in developed countries that is wasted – simply thrown away.

27

The percentage of journeys taken in the Netherlands on a bicycle (this compares to just 2% of journeys in the UK).

27

Despite being over 27 times smaller, Norway's total coastline is longer than the US's.

28

The percentage of the Earth's surface covered by the Pacific Ocean.

30

The number of people who speak the Cambap language of Central Cameroon.

30

The percentage of Chinese adults who live with their parents.

30

The percentage of degrees awarded in India for engineering.

32

Luxembourg has more immigrants – 32% of the population – than any country in the world.

32

The number of mountain peaks above 4,000 metres in Switzerland.

33.3

The percentage of the world's oranges grown in Brazil, which is also the world's largest producer of orange juice. However, you won't always see Brazil on the carton because they export the concentrate to other countries, which then claim the juice as their own.

35

The percentage of Australians who are of Irish origins.

37

The length in metres of the D River in Lincoln City, Oregon – the shortest river in the world.

38

The percentage of North America that is wilderness.

38

The steepest street in the world is Baldwin Street in Dunedin, New Zealand which has an incline of 38%.

45

The width in centimetres of a street in Italy.

46

The percentage of the world's water that is in the Pacific Ocean.

47

The number of sounds in the Japanese language.

48

The number of kilometres by which the Sahara desert expands annually.

50

The percentage of
the world's cork that
comes from Portugal.

50

The percentage
of 15-year-olds in
Greenland who smoke.

50

The percentage of his
or her annual holiday
entitlement that the
average Japanese
worker takes.

50

The percentage of the
world's population that has
never made or received
a single telephone call.

50

The percentage of
motorbike accidents that
happen at the weekend.

51

The number of cars per
annum that overshoot
and drive into the
canals of Amsterdam.

52

Switzerland has the
best recycling rate in
the world - recycling
52% of all its waste.

58

The percentage of Turkish
people who admit to
having had an extra-
marital affair. The others in
the top 5 are all Nordic.

59

Litres of wine the average
French person drinks

every year. The average Russian drinks ninety bottles of vodka.

60

The Amayra guides of Bolivia are said to be able to keep pace with a trotting horse for over 60 miles.

60

The number of Turkish companies that manufacture chewing gum.

61

The number of square miles that comprise the country of Liechtenstein.

66

The percentage of prisoners in Swiss jails who are non-Swiss nationals.

70

The percentage of computer-virus writers

who work for organized crime syndicates.

70

Bulgaria produces about 70% of the world's attar of roses – an ingredient in the most expensive perfumes. Two thousand petals are needed to make a single gram of attar of roses

75

The percentage of all the countries in the world that are north of the Equator.

75

There is no point in England more than 75 miles from the sea.

80

The percentage of the world's opals produced by the Australian town of Coober Pedy.

81

The number of provinces in Turkey.

81.5

The percentage of rented housing in Estonia, the country with world's most rented housing.

82

The percentage of Australians who gamble regularly.

84

The percentage of Botswana that's covered by the Kalahari Desert.

85

The percentage of international telephone calls that are conducted in English.

85

The number of letters in Taumatawhakatangihangakoauauotamateaturipukakapikimaungahoronukupokaiwhenuakitanatahu – the name of a hill in the Hawke's Bay region of the North Island of New Zealand. This is the longest geographical name in the world. It translates into English (from Maori) as 'The summit where Tamatea, the man with the big knees, the climber of mountains, the landswallower who travelled about, played his nose flute to his loved one'. The locals call it Taumata.

90

The percentage of Canada's citizens who live within 100 miles of the US border.

90
The percentage of the world's amber production that comes from the Sambia peninsula on the Baltic Sea.

93
The percentage of cremations in Japan, compared with Great Britain at 67% and China at 49%.

96
The number of metropolitan districts in France.

100
The ambulance and firefighter telephone number in Belgium.

150
The number of miles that separate Wallis and Futuna in the central South Pacific. Since 1959, they have been joined together as a French overseas territory.

158
The number of verses in the national anthem of Greece.

160
The number of cars that can drive side by side on the Monumental Axis in Brazil, the world's widest road.

171
the length in kilometres of the Suez Canal.

192
The number of deaths for every thousand births in Angola, giving it the worst infant mortality rate in the world.

200
The world's average school year. In the US, it is 180 days; in Sweden 170 days. In Japan, however, the school year is 243 days long.

294
The number of steps in the Leaning Tower of Pisa.

300
There are 300 times as many sheep as there are people in the Falkland Islands.

400
The surface of the Dead Sea is 400 metres below the surface of the Mediterranean Sea.

441
The number of Australian airports with scheduled flights.

600
The number of bicycles for every car in China.

780
The land size in square kilometres of King Fahd International Airport, in Saudi Arabia. It is larger than the country of Bahrain.

820
The number of languages spoken in Papua New Guinea.

1,000
The number of people who are caned every

year in Singapore. The bamboo cane is soaked in water overnight to prevent it from splitting.

1,040
The number of islands around Britain, one of which is the smallest island in the world: Bishop's Rock.

1,788
The number of rooms in the Palace of the Sultan of Brunei – the largest residence in the world.

1,792
The number of steps in the Eiffel Tower.

2,334
The number of kilometres that Tristan de Cunha – the most remote country or settlement in the world – is from its nearest neighbour.

2,861
The average number of cigarettes a year smoked by a Serbian adult (the most in the world).

3,000
Up to 3,000 species of trees have been catalogued in a single square mile of the Amazon jungle.

3,658
If the Earth were smooth, the ocean would cover the entire surface to a depth of 3,658 metres

3,808
The average number of calories-consumed per person per day in Denmark, making it the world's top calorie consuming country. The Americans come second with 3,700 a day. The average across the world is 2,745.

4,132
The length of the River Nile in miles.

5,180
The highest home in the world is a shepherd's hut in the Andes at 5,180 metres.

6,000
The estimated temperature at the centre of the Earth.

6,000
Duration in years of the Guinness Brewery's Dublin lease.

6,852
The number of islands that make up Japan. However, just four – Hokkaido, Kyushu, Shikoku and Honshu – make up 97 per cent of the land area.

11,035
The depth in metres of the deepest part of the ocean – the Mariana Trench, which is located south-east of Japan. At this depth, an iron ball would take more than an hour to sink to the bottom. By comparison, the height of Mount Everest is 8,848 metres.

14,000
The number of daily trains run by the Indian railway system, which transports over six billion passengers each year.

18,000
The number of discarded glass bottles used to make a house in Canada.

20,000
The number of pupils in the world's largest school (in the Philippines).

27,000
The age of Dolní Věstonice, a small village in the South Moravian Region of the Czech Republic, making it the most ancient village in the world.

30,000
The length in miles of the world's longest road, the Pan-American Highway which stretches from Alaska to Argentina (except for a 54-mile section of rainforest).

40,000
The number of statues of Joan of Arc in France.

67,807
The estimated number of hospitals in China, the largest in the world.

179,584
The number of islands in Finland, more than any other country.

187,880
The number of lakes in Finland.

500,000
The number of semi-automatic machine guns in Swiss homes.

845,000
The number of dams in the world.

2,500,000
The number of rivets in the Eiffel Tower.

3,000,000
The pressure at the Earth's inner core is 3,000,000 times Earth's atmospheric pressure.

6,000,000
The number of trees in the Forest of Martyrs near Jerusalem, symbolizing the Jewish death toll in World War Two.

7,692,024
The size of Australia in square kilometres.

10,180,000
The size of Europe in square kilometres.

13,500,000
The size of Antarctica in square kilometres.

16,000,000
The number of bicycles in the Netherlands (which has more bicycles per person than any other country in the world).

29,800,000
The size of Africa in square kilometres.

43,998,000
The size of Asia in square kilometres.

70,000,000
The number of people across the globe who have Irish ancestors.

197,000,000
The surface area of the
Earth in square miles .

200,000,000
The approximate age
of the Atlantic Ocean,
making it the youngest
of the world's oceans.

800,000,000
The number of adults in
the world who don't know
how to read or write.

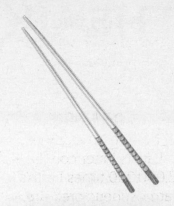

45,000,000,000
The number of chopsticks
used annually in China.

**6,588,000,000,
000,000,000**
The Earth's approximate
weight in tons.

THE UK

0
The number of women's prisons in Wales (female Welsh offenders are sent to prisons in England).

0.16
The shortest distance (in miles) between stations on the London Underground. It's between Leicester Square and Covent Garden on the Piccadilly line and, taking into account the stairs etc., it's much quicker to walk.

0.9
The percentage of the land mass of Great Britain that is covered by roads.

1
The number of British-born Popes: Adrian IV (born Nicholas Breakspeare) 1154–9.

1
The number of shops that sold soft toilet paper when it first went on sale in Britain in 1947 (Harrods).

1
The number of people who have been born on the London Underground in a tube carriage (Thelma Ursula Beatrice Eleanor was born in 1924 on a Bakerloo line train at Elephant & Castle. Her real name was Marie Cordery).

1
The number of daffodils

the Isles of Scilly Wildlife Trust pays Prince Charles annually in rent.

....

1
The number of British counties with two coasts (Devon).

....

1
The number of days that Labour's Alfred Dobbs served as an MP (he died in 1945).

....

2
The number of people whose coffins were transported on the London Underground (William E. Gladstone and Dr Barnado).

....

2
The number of months that a road in Charlcombe, Somerset is closed for every spring to allow frogs and toads to cross safely.

2
The number of stations on the London Underground which have all five vowels in them (Mansion House and South Ealing).

....

2.2
The number of healthy pregnancies that the average British woman will have in her lifetime (but she'll give birth to only 1.7 children – the difference is because of the number of abortions).

2.5
The height in metres of the London Underground's shortest lift shaft (King's Cross).

....

2.8
The number of unhealthy snacks the average British person eats each day.

3

The number of men named Richard Fort who represented Clitheroe in the House of Commons.

4

The number of women awarded the George Cross for acts of bravery during World War II, out of a total of 406.

4

The number of times the word 'Queen' occurs in the first verse of the National Anthem.

4

The number of years the Army Cyclist Corps, which operated the British Army's bicycle infantry, existed.

4

The percentage of British house vendors who admit to removing the house number when they leave.

4

British men commit four times as many crimes as British women.

4

The percentage of Britons who believe in the Loch Ness Monster.

4

The number of public toilets in the whole of London, England, in 1358. The largest, on London Bridge, emptied straight into the River Thames.

5

The number of pounds Harry Beck was paid in 1933 to design the map of the London Underground.

6

Heathrow Airport is six times the size of the country of Monaco.

7

The percentage of British adults who go to the opera at least once a year.

7

The average British adult moves home every seven years.

7

In Britain, someone with the surname Patel is seven times more likely to be a multi-millionaire than someone with the surname Smith.

7

The percentage of British people who say that they trust their neighbours.

7.5

The number of units of alcohol the average British woman drinks each week. A French woman drinks 9.

8

The UK is eight times bigger than Belgium. Saudi Arabia is eight times bigger than the UK.

10

The number of platforms Moorgate and Baker Street underground stations each have.

10

The percentage of British adults who admit to wearing the same item of underwear three days in a row.

10

British men are ten times as likely as British women to be convicted of dangerous driving offences.

11

The number of kilograms of chocolate consumed per person per year in the UK.

11.5

The duration in years of the average British marriage. After thirteen years, the marriage is more likely to end because of death rather than divorce. The average length of a co-habiting relationship is three years.

12

The percentage of Britons who believe that they have seen ghosts.

14

The percentage of (adult) Britons who have never got drunk.

15

The number of National Parks.

Dartmoor (designated a National Park in 1951)
•
The Lake District (1951)
•
The Peak District (1951)
•
Snowdonia (1951)
•
Pembrokeshire Coast (1952)
•
North Yorkshire Moors (1952)
•
Exmoor (1954)
•
The Yorkshire Dales (1954)
•
Northumberland (1956)
•

The Brecon Beacons
(1957)

•

The Broads (1989)

•

Loch Lomond and The
Trossachs (2002)

•

The Cairngorms (2003)

•

The New Forest (2005)

•

The South Downs (2009)

15

The number of hours a
year the average Briton
devotes to Christmas
shopping. Around
25% of that time is
spent in queues.

16

The number of tunnels
beneath the River Thames
– mostly for London
Underground trains.

17

The number of units of
alcohol the average British
man drinks each week.

19

The percentage of British
people who believe
in reincarnation.

19

The percentage of British
people who've never
visited a McDonald's.

21

The number of square
miles of rural Britain that is
built over every single year.

21

The percentage of
British adults who have
no savings at all.

22

Birmingham has 22 more
miles of canal than Venice.

23
The percentage of the UK population that lives in the countryside.

26
The percentage of Britons who believe in UFOs.

27
The percentage of Britons who have never read Shakespeare.

29
The percentage of British families that sit down to eat a meal together more than once a week.

29
The number of stations on the London Underground that are south of the Thames.

29
The number of weeks the average person keeps magazines before throwing them out.

29
The percentage of British women who cut off the size labels from their clothes.

31
The number of times the UK would fit into Australia.

33
The number of litres of bottled water the average Briton drinks in a year.

36
The number of miles travelled by taxi by the average Briton in a year.

36
The percentage of Britons who have never been to a football match.

37
The weight in kilograms of chips eaten by the average Briton in a year.

40
The percentage of *Woman's Hour* listeners who are men.

42
The distance in kilometres from Dover to Calais.

45
The percentage of Britons who reckon they don't get enough sleep.

45
The percentage of the London Underground network that is actually underground.

49
The average age of the first-time British grandparent.

50
Britain's first Indian restaurant was opened more than 50 years before the first fish-and-chip restaurant.

50
The number of storeys in One Canada Square, the building commonly referred to as Canary Wharf.

55
The percentage of British workers who say that they never take a lunch break.

55.2
The height in metres of the London Underground's deepest lift shaft (Hampstead).

59

The percentage of Londoners who give/receive a daily kiss to/from their partner (the percentage of Scots is 38).

62

The percentage of British people who speak no other language than English.

62

The number of hours a week the average mother spends on household tasks in the UK.

67

The percentage of all the land in England that is owned by less than one per cent of the people.

68

The percentage of Britons who say they believe in the existence of ghosts and/or spirits.

70

The number of times the UK would fit into Russia.

70

The percentage of British divorces that are instigated by women.

70

The percentage of British homes that are owner-occupied.

75

The percentage of Britain's rural parishes that have no daily bus service.

75

The percentage of British drivers who sing along to the car stereo.

75

The percentage of British men who prefer a shower to a bath (women split 50/50).

84

The number of livery companies in London.

85

The number of cash-machine transactions that take place every second in the UK.

115

The size of Belfast in square kilometres.

117

The length in kilometres of Hadrian's Wall.

140

The size of Cardiff in square kilometres.

158

The height in metres of Blackpool Tower.

189

The number of MPs 'elected' in the 1832 general election who were returned without any vote being held.

250

The number of people who are killed or injured on motorway hard shoulders in Britain every year.

250

The number of miles of railways on the London Underground system.

260

The number of people per square kilometre in Britain (there are 3 per square kilometre in Australia).

264

The size of Edinburgh in square kilometres.

287

The number of stations on the London Underground.

288

The number of years it took for the most overdue library book to be returned. Robert Walpole, the father of *the* Robert Walpole who was Britain's first Prime Minister, borrowed a book from Sidney Sussex College, Cambridge, in 1667. It was returned in 1955.

300

The English Channel grows about 300 millimetres each year.

345

The UK is 345 times bigger than Singapore.

400

The average Briton spends more than 400 hours a year shopping.

472

The number of beaches in the UK.

528
The number of steps
leading to the top of St
Paul's Cathedral.

850
The number of Christmas
cards sent by the
Queen and the Duke of
Edinburgh annually.

1,572
The size of London in
square kilometres.

2,000
The number of Cornish
speakers in Britain.

3,000
The number of meals
of spaghetti bolognese
the average Briton eats
in a lifetime.

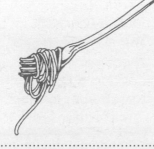

4,000
The number of people
treated in British hospitals
every week as a result of
DIY repairs going wrong.

6,289
The number of islands
that make up the British
Isles (most of them off
the coast of Scotland).

10,911
The number of miles
of railway in the UK.

35,000
The number of biscuits
the average Briton will
eat in a lifetime.

35,000
The number of hectares
of the UK devoted
to growing peas.

60,000
The number of British children who are educated at home.

80,000
The number of times a day the emergency number, 999, is called.

92,900
The amount of glass in square metres that the Crystal Palace – built for the Great Exhibition of 1851 – contained.

150,000
The number of Britons who work on the land – compared to more than two million a hundred years ago.

250,000
The number of roads in Britain. If stretched end to end, they would reach the Moon.

500,000
The number of listed buildings in the UK.

5,900,000
The number of CCTV cameras in Britain (one for every 14 people).

600,000
The number of vegans in the UK.

4,500,000
The number of gym members in the UK (but only a quarter of that number exercise on a regular basis).

8,000,000
The number of
Christmas trees sold
in Britain each year.

8,000,000
The number of disposable
nappies thrown away
in the UK every day.

11,000,000
The number of cows
in the UK.

13,000,000
There are some 13 million
more credit cards in Britain
than there are people.

14,000,000
The number of tins of
sardines – the equivalent
of 5,400 tons – sold
annually in Britain.

75,000,000
The number of cups of
coffee drunk by Britons
each day.

150,000,000
The number of avocados
sold in the UK every year.

300,000,000
The amount in pounds
sterling that women spend
each year on removing
unwanted body hair.

400,000,000
The number of
diagnoses made annually
by British GPs.

1,500,000,000
The number of loo
rolls Britons use in
total each year.

10,000,000,000
The number of eggs
consumed every
year in Britain.

THE USA

0

The number of wars in which mules were used that the US has lost.

0

The number of US green cards that were actually green between 1964 and 2010, when the colour was reintroduced.

0

The number of times the letter 'q' appears in the names of any of the fifty States of the US.

0

The number of US presidents who have been only children.

0

The number of former US presidents to have died in the month of May.

1

The number of US states where adults don't have to wear a car seat belt (New Hampshire).

1

The number of mainland US states that border on just one other state (Maine).

1

The number of US presidents to deliver an inaugural address without using the word 'I'. (Theodore Roosevelt. Note that Abraham Lincoln, Franklin D. Roosevelt and Dwight D. Eisenhower all used 'I' just once in their inaugural addresses.)

1

The number of US presidents to have had Parisian Metro stations named after them (Franklin D. Roosevelt).

1

The number of US presidents who were bachelors (James Buchanan).

1

The number of US presidents never to have been elected as either president or vice-president (Gerald Ford).

1

The number of Americans to get a favourable mention in Hitler's autobiography, *Mein Kampf* (car manufacturer Henry Ford).

1

The number of US states that permit residents to cast ballots from space (Texas).

1

The number of US states that can be typed on one row of keys on a typewriter/computer keyboard (Alaska – the middle row).

1

The number of men who were president of the US for just a day (in 1849, David Atchison, after becoming president of the United States, spent most of the day sleeping).

2

The number of emails sent by Bill Clinton during his eight-year presidency (one was to John Glenn aboard the Space Shuttle, and the other was to test the email system).

2

The number of lawsuits in the US every minute.

2

The number of cars in the state of Ohio in 1895 (yet somehow, they managed to collide).

2

The number of credit cards for every person in the US.

2

The number of former US presidents who died on 4 July (Independence Day) 1826 (John Adams (the second President) and Thomas Jefferson (the third) who both died on the very same day).

2

The number of Southern state capitals not occupied by Northern troops during the American Civil War (Austin, Texas, and Tallahassee, Florida).

4

Every US bill – regardless of denomination – costs four cents to make.

4

The number of time zones covered by the University of Alaska.

4

The number of states where the first letter of the capital city is the same letter as the first letter of the state: Dover, Delaware; Honolulu, Hawaii; Indianapolis, Indiana; and Oklahoma City, Oklahoma.

2

Ulysses S. Grant was tone-deaf and once said: 'I only know two tunes, one of them is *Yankee Doodle* and the other isn't.'

3

The number of US states' names that begin with double consonants (Florida, Wyoming and Rhode Island).

3

The minimun number of people who can be accused of rioting (in the UK, there have to be at least twelve people to constitute a riot).

5

The central plaza of the US Pentagon covers five acres and is the largest 'no-salute, no-cover' area in the world (outside of combat zones): military personnel don't have to wear caps or salute superior officers.

5

The number of states you can see from the top of the Empire State Building on a clear day (New York, New Jersey, Connecticut, Massachusetts and Pennsylvania).

6

The number of months it took in 1938 for Gatke Hall, a former post office, to be moved completely intact on rollers down a city street to its new home at Willamette University in Salem, Oregon.

6

In 1986, Nathan Hicks of St Louis, Missouri, shot his brother Herbert dead because he used six toilet rolls in two days.

6

In the state of Michigan, no one is more than six miles from a lake.

6

The percentage of American men who propose marriage by phone.

7

The percentage of the US population that eats at McDonald's daily.

7

The number of hills on which Seattle, Washington, was built.

9

The number of milligrams of rat droppings per kilo of wheat deemed permissible by the US Food and Drug Administration (FDA).

9

In just one day, Americans use enough toilet paper to wrap around the world nine times. If all the toilet paper they used were on one giant roll, they would be unrolling it at the rate of 7,600 miles per hour).

9.4

The number of years that the average American marriage lasts.

10

The number of cows there are for every person in the US state of Vermont.

11

The number of states in the Confederacy.

12

The number of 'filthy' words laid down by the Federal Communications Commission in 1973, all of which are still widely used today.

12

The percentage of Americans who think that Joan of Arc was Noah's wife.

14

The number of impulse decisions made by American supermarket shoppers in a single trip to the supermarket.

16

When Calvin Coolidge was being driven in a car, he would always insist that the driver didn't exceed 16 mph.

17
The number of tons of gold used annually to make wedding rings in the US.

17.5
The Pentagon has 17.5 miles of corridors.

18
Americans eat 18 acres of pizza every day.

18
The percentage of American pet owners who share their beds with their pets.

19
The number of guns in a salute for the vice-president of the United States.

20
The percentage of Americans who think that the sun orbits around the Earth.

20.4
The number of gallons of milk drunk by the average American in a year.

21
The number of guns in a salute for the President of the United States.

21

Alaska is big enough to hold the 21 smallest US states.

24

The number of dollars paid to the Native Americans for Manhattan Island.

25

The percentage of businesses in the US that are franchises.

25

The number of people per square kilometre in the US (there are 351 people per square kilometre in Japan).

25

The percentage of the world's prison population incarcerated in the US (which has just five percent of the world's population).

26

The number of US states with an official state reptile.

27

The number of World War II US Marines who threw themselves on to exploding grenades to save their comrades. Only three of the 27 survived.

28

The number of pigs a typical American eats in their lifetime.

30

The US Food and Drug Administration (FDA) allows 30 insect fragments (and one or more rodent hairs) per 100 grams of peanut butter.

30

A US judge once ordered a family to eat dinner together for 30 days (he even sent a parole officer along to verify that they did so).

30

The percentage of all the nonbiodegradable rubbish buried in American landfills that is made up of disposable nappies.

33

The percentage of Americans who flush the toilet while they're still sitting on it.

33

The percentage of American dog owners who admit that they talk to their dogs on the phone or leave messages for them on an answering machine when they're away.

37

The number of US states in which it is mandatory to wear a helmet while cycling.

38

Utah's State Route 69 was renamed State Route 38 because people were stealing road signs (because of the sexual connotations of the number 69).

39

The percentage of Americans who have passports. Sixty per cent of Canadians have passports but only 1.5% of Chinese people do.

42

In 1935, the police in Atlantic City, New Jersey, arrested 42 men on the beach. They were cracking down on topless bathing suits worn by men.

47

Ohio is listed as the 17th state in the US, but technically it is number 47. Until 7 August, 1953, Congress forgot to vote on a resolution to admit Ohio to the Union.

50

The percentage of Americans who live within 50 miles of their birthplace.

50

The percentage of people in Kentucky who are teenagers when they get married for the first time.

50

The percentage of American women who say they would marry the same man if they had to do it all over again. (The percentage of men is 80.)

50

The percentage of all marshmallows consumed in the US that have been toasted.

55

All the earthworms in America weigh 55 times what all the people weigh.

60

The number of towns in the US where the word turkey appears in the name.

67

The number of hours each year people in Washington DC spend sitting in traffic. The average is 52 hours.

70

The percentage of the world's lawyers who practise in the US.

71

The percentage of American office workers who, when stopped on the street for a survey, agreed to give up their computer passwords in exchange for a chocolate bar.

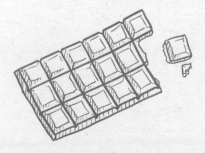

72

The percentage of Americans who sign their pets' names on greetings cards they send out.

75

The percentage of US citizens who live on just two per cent of the land.

90

The percentage of New York taxi drivers who were born outside of the US.

97

The percentage of all paper money in the US which contain traces of cocaine.

99

The percentage of the pumpkins sold in the US that end up as jack-o'-lanterns.

100

The number of Florida Black Bears that are killed on Florida roads each year.

120

The amount of meat in kilograms that the average American eats per year. The average per person worldwide is 42kg.

132

The number of rooms in the White House.

132

The number of islands that make up Hawaii.

184

The number of people bitten by rats in New York in an average year.

190

The number of pieces of gum the average American chews each year.

217

The number of times an eleven-year-old American boy farted in five minutes on a radio call-in show.

311

The number of languages spoken in the US.

365

The number of Americans who drown in their own baths in an average year. More Americans, however, simply freeze to death (just under twice as many).

538

New York City has 570 miles of shoreline.

556
The number of officially recognized Native American tribes.

765
The number of cars per thousand people in the US (the most in the world).

1,600
The number of people bitten by other people in New York City in an average year.

5,000
The number of people who, when the first ballpoint pens went on sale in New York in 1945, queued to buy one for $12.50.

5,400
The number of spoons in the New Jersey spoon museum.

8,000
The number of US troops still listed as missing in action from the Korean War.

12,500
The number of puppies born every hour in the United States.

62,654
The number of flying hours achieved by John Edward Long. From 1933 to 1977, he spent more than seven years' airborne.

200,000
The number of members of the American Nazi Party in 1939.

425,000
The number of German
prisoners of war held in
700 camps throughout the
US during World War II.

14,000,000
The number of cars
scrapped annually
in the US.

72,000,000
The number of books in
the Library of Congress
in Washington, DC (the
largest library in the world).

HUMANS BEING

0.1
Every time you lick a stamp, you're consuming one-tenth of a calorie.

0.7
At any one time, 0.7% of the world's population is intoxicated.

1
We all weigh one per cent less at the equator.

1.36
The weight in kilograms of the largest known kidney stone.

1.7
The number of deaths in the world every second.

2
The number of primary emotions in decision-making.

2
The number of metres that dentists recommend a toothbrush be kept away from a toilet to avoid airborne particles resulting from the flush.

2
The number of teeth lost, on average, every ten years by a person who smokes twenty cigarettes a day.

2

The number of ballpoint pens lost each week by the average person.

2

Happy events – like family celebrations or evenings with friends – boost the immune system for the following two days.

2

The approximate number of years the average person spends in the course of their lifetime on the phone.

2.5

The number of centimetres the average single man is shorter than the average married man.

2.72

The weight, in kilograms, of lipstick the average woman uses during her lifetime.

3

The number of people in the world with the blood type A-H.

3

The number of times around the world a person will walk in the average lifetime.

3.5

The number of calories burnt up by laughing for two minutes.

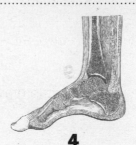

4

The number of times women are more likely to have foot problems than men.

4

The percentage of babies born on their actual due date.

4

A household headed by someone under 25 is four times more likely to be burgled than one headed by someone over 65.

4

The number of times men are more likely to sleep naked than women.

4

The number of gallons of water wasted each time you brush your teeth with the tap running.

4.9

A healthy adult can draw in up to 4.9 litres of air in a single breath, but at rest only about 5% of this volume is used.

5

The approximate number of years the average person will spend dreaming during their lifetime.

5

The percentage of men who admit to having used their toothbrush after it fell down the toilet.

5

The number of minutes the average bout of hiccups lasts.

5.6

The percentage of people who suffer from restless-leg syndrome.

6

The number of months in a lifetime that the average person will spend on the toilet.

6

The percentage of women who fail to cry at least once a month.

6

Grapefruit scent makes middle-aged women seem six years younger to men (it doesn't work the other way round).

7

The average person falls asleep in seven minutes.

7

There's a systematic lull in conversation every seven minutes.

7

The average person throws away seven times their body weight in rubbish every year.

7

The number of times a minute the average computer user blinks.

7.5

On average, murderers are 7.5 years younger than their victims.

7.5

The world record for holding one's breath is 7.5 minutes.

7.5

The number of miles an office chair with wheels will travel in a year.

8

The number of houses the average house buyer looks at before buying one.

8.5
The average clean-shaven man will remove 8.5 metres of hair in his lifetime.

10
The number of days a person can go without *any* sleep before dying. A person will die from total lack of sleep sooner than from starvation.

10
There are ten distinct types of laughter (amused laugh, joy laugh, sympathetic laugh, polite laugh, relief laugh, disappointment laugh, embarrassed laugh, stressed laugh, comment laugh, ironical laugh).

10
Women are ten times less likely than men to suffer from colour blindness.

10
The percentage of secretaries who admit to having been romantically involved with their bosses.

10
Over the past 150 years the average height of people in industrialized nations has increased by ten centimetres.

10
People who smoke have ten times as many wrinkles as people who don't smoke.

12

The number of times each day that newborn babies are given to the wrong mother in maternity wards across the world.

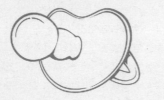

12

The percentage of married British men who claim to do most of the housework, compared to thirty in Norway and forty in China.

12.5

The percentage of boss–secretary romances which end in marriage.

13.7

The temperature in Celsius of Sweden's Anna Bågenholm when she got trapped under a layer of ice in a river for eighty minutes – the lowest survived body temperature ever recorded in a human.

15

On average, women sleep about 15 minutes longer a night than men do.

17

The number of times a day the average person laughs.

18

The average time – in minutes – spent looking at a property before buying.

20

The percentage by which your IQ can drop after a three-week holiday.

20

The percentage of people in the whole history of

mankind to have lived beyond the age of 65 who are alive today.

20
You burn 20 calories an hour chewing gum.

21
On average, left-handed graduates earn 21 percent more than right-handed graduates.

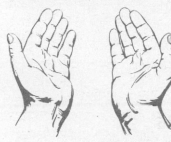

22
The number of times a day the average person opens the fridge.

22
The percentage of twins who are left-handed – more than twice as many as non-twins.

22
The average time – in minutes – spent looking at an expensive item of clothing before buying.

25
The percentage of New Year's resolutions that are broken within the first week of January.

28
On average, the number of minutes per day that children between the ages of 2 and 7 spend colouring.

30
The average person's eyes will be closed about 30 minutes a day due to blinking.

33.3
The percentage of cancers that are sun related.

33.3
The percentage of the world's population that is under 18.

35
The number of centimetres apart we stand when we converse.

37
The percentage of men who shut their eyes during a kiss, compared to 97 per cent of women.

39
A really tongue-twisting kissing session exercises 39 different facial muscles and can burn up 150 calories – more than a 15-minute swim or a hill climb carrying a 20 kg rucksack. An ordinary peck uses up just three calories.

40
The percentage of people who come to a party in your home who have a look in your medicine cabinet.

40
The percentage of women who have hurled footwear at a man.

40
The percentage of Christmas presents bought on the internet.

50
The percentage of men who fail to cry at least once a month.

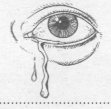

50
The percentage of taste buds that most people lose by the time they reach the age of sixty.

55
The percentage of men who wash their hands after leaving a public toilet.

56
The average person's left hand does 56% of the typing.

60
The percentage of people who answer 'red' when asked to name a colour.

64
The percentage of women who sleep on the left side of the bed.

64
The percentage of people who can roll their tongue.

67
The percentage of the world's illiterate adults who are women.

70
The number of decibels in a really loud snore (the loudest recorded is 92, which is almost as loud as the noise of a pneumatic drill).

72
The number of minutes for which the average man can shop before becoming bored (the average woman can last for a hundred minutes).

77
The number of minutes on an average day the average adult spends eating.

80

The percentage of millionaires who drive second-hand cars.

80

The percentage of women who wash their hands after using a toilet.

80

The percentage of what we learn on any given day that we forget.

86

The percentage of women who look at price tags when they shop (that figure goes down to 72% for men).

90

The percentage of people who have the disease Lupus who are female.

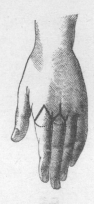

90

The percentage of amputees who report retaining all or part of the missing limb in their body schema (their sense of selves).

92

The percentage of couples who stick to the same side of the bed every night.

95

The percentage accuracy with which a person's gender can be guessed just by smelling their breath.

97

The percentage of people who, when offered a new

pen to write with, will write their own name.

98
The percentage of men who, when left alone with a tea cosy, will put it on their heads.

100
The number of people who choke to death on ballpoint pens every year.

102
The record number of days for a person having constipation.

113
The weight in kilograms of a five-year-old girl who wasn't able to sweat. She was put on display at an exhibition in Vienna in 1894.

113
The number of minutes a newborn baby cries, on average, every day.

116
One person in two billion will live to be 116 or older.

140
For every mile travelled, death is 140 times more likely in a car than in a plane. However, cars are actually safer than planes on a per-journey basis.

145
The number of days in a lifetime that the average man will spend shaving.

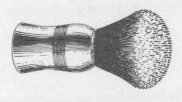

150
The number of days it takes for a fingernail to grow from cuticle to fingertip.

200
Babies born in May are on average 200 grams heavier than babies born in other months.

275
The number of words the average reader can read per minute.

300
The number of times a day six-year-olds laugh.

300
The approximate number of microscopic saliva droplets the average talker sprays per minute. About 2.5 droplets per word.

400
The number of questions the average four-year-old child asks a day.

700
Wearing headphones for an hour increases the bacteria in your ears 700 times.

978
The number of consecutive days that the longest-recorded sneezing fit lasted.

1,140
The approximate number of phone calls the average person makes each year.

4,500
The number of steps the average person takes in a day.

5,500
The number of words in the average adult's vocabulary.

5,800
The number of nappies used by the average baby.

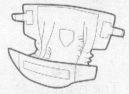

6,500
The chances of being injured by a toilet seat at some point in your lifetime are one in 6,500.

7,000
The number of words a day the average man speaks (although almost 80 per cent is self-talk – i.e. talking to oneself). Women use 20,000 words a day.

7,163
The number of baths the typical person will have in a lifetime.

12,568
In 1967, a man had 12,568 boils on his body. No part of his skin was left uncovered by boils. The largest boil measured was nine centimetres in diameter. That's just three centimetres smaller than a CD.

23,040
The number of times the average person breathes in a day.

100,000
The number of miles the average person walks by the time they reach the age of 85.

200,000
The number of frowns
it takes to make a
permanent wrinkle.

220,000
The number of hours
the average person
sleeps in a lifetime.

55,000,000
The number of gallons
of blood a human heart
pumps by the age of 70.

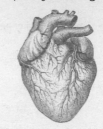

THE HUMAN BODY

0
The number of kneecaps a baby has. They don't get them until the child reaches two (at the earliest).

0.01
Parasites count for 0.01% of our body weight.

0.02
When you stub your toe, your brain registers pain in 0.02 of a second.

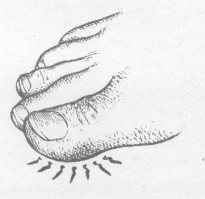

0.05
The breadth of a typical hair in centimetres.

0.6
The average person is 0.6 of a centimetre taller at night.

1
The number of minutes it takes for blood to travel through the whole human body.

1

The number of primates that don't have pigment in the palms of their hands (humans).

1.13

The mouth produces 1.13 litres of saliva a day.

1.36

The weight in kilograms of the human brain – most of it water.

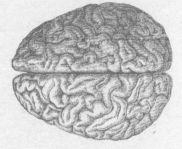

2

The amount in metres of nose hair an average person grows.

2

Your stomach has to produce a new layer of mucus every two weeks otherwise it will digest itself.

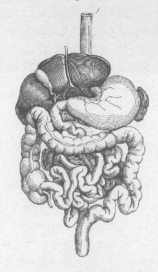

2.8

The amount of sweat in litres an average man on an average day excretes.

3

The amniotic fluid that surrounds a baby in the womb is replaced completely every three hours.

3

The weight, in grams, that one human hair can support.

3

The number of colour receptors we have in our eyes (goldfish have four, and mantis shrimp have ten).

· · · · · · · · · · · · · · · ·

3

The number of layers of human skin.

· · · · · · · · · · · · · · · ·

4

The approximate weight in kilograms of ashes of the average cremated person.

· · · · · · · · · · · · · · · ·

4

Fingernails grow four times faster than toenails.

· · · · · · · · · · · · · · · ·

4

The number of senses stronger than taste, the weakest of the five senses.

· · · · · · · · · · · · · · · ·

5

The number of months the average eyelash lasts.

· · · · · · · · · · · · · · · ·

6

Until babies are six months old, they can breathe and swallow at the same time.

· · · · · · · · · · · · · · · ·

6.8

The number of kilograms of hydrogen in the average human body.

· · · · · · · · · · · · · · · ·

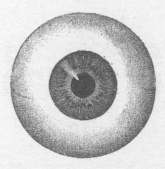

7

The weight in grams of an eyeball.

· · · · · · · · · · · · · · · ·

7

The human body grows the equivalent of a new skeleton every seven years.

· · · · · · · · · · · · · · · ·

8

The percentage of the body's weight that consists of blood.

8

The number of muscles in the tongue.

9

The number of metres the human heart has enough pumping pressure to squirt blood.

9

The length in metres of the digestive tract.

10

The percentage by which the male human brain is heavier than the female brain (men also have on average ten per cent more red blood cells than women).

10

The number of human body parts that are only three letters long: eye, ear, leg, arm, jaw, gum, toe, lip, hip and rib.

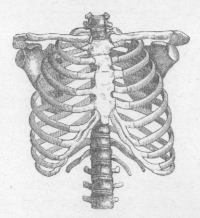

12

The number of days it takes for your body to generate a whole new set of taste buds.

12.7

The length in centimetres the average beard grows in a year.

15

The number of watts of electricity it would take to stop the heart.

15
The loss of 15% of the body's water can be fatal.

16
The percentage of our bodyweight accounted for by our skin.

17
The number of hip muscles that work together in complex ways to produce a wide range of movement.

17
The number of muscles we use to smile.

17
The length in centimetres of the sternum.

18.5
The length in centimetres of the innominate (the bone linking the hip and the pelvis).

20
The percentage of all the body's available oxygen used by the kidneys.

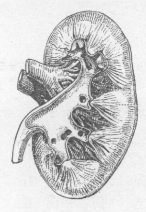

20
The number of watts of power that the brain runs on (about enough to power a very dim light-bulb).

20
The number of seconds it takes for a red blood cell to circle the whole body.

25
If all 600 muscles in your body pulled in one direction, you could lift 25 tons.

25
The percentage of a newborn baby's weight accounted for by its head.

25
The number of muscles we use to swallow.

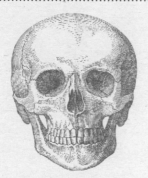

29
The number of bones in the human skull.

30
The length in metres of all the eyelashes shed by a human in their life.

36.46
The length in centimetres of the average humerus.

40
The percentage by which men sweat more than women.

40
The number of kilograms of dead skin we shed in a lifetime.

40.5
The length in centimetres of the average fibula.

43
The number of muscles we use to frown.

43.03
The length in centimetres of the average tibia.

46
Humans have 46 chromosomes (peas have 14 and crayfish have 200).

46
The number of miles
of nerves in the adult
human body.

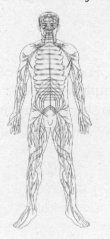

48
If your body's natural
defences failed, the
bacteria in your gut
would consume you
within 48 hours.

48
The average length,
in centimetres, of the
femur in an adult male.

54
The number of muscles
we use every time
we step forward.

56
The number of days
it takes for a plucked
hair to reappear.

65
The number of hairs
shed by the average
person each day.

72
The number of different
muscles it takes to
produce human speech.

98
The percentage of the
atoms in our bodies that
are replaced annually.

100
How fast a human sneeze
travels in miles per hour.

121
The number of litres of
tears an individual sheds
in a lifetime.

170

The speed, in metres per second, at which pain travels through our bodies.

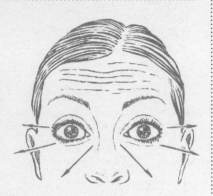

200

The angle in degrees of the average person's field of vision.

206

The number of bones in a human adult. At birth, an infant has 350 bones. As the child grows, many bones fuse with other bones.

240

The number of gallons of air we breathe every hour.

250

The number of sweat glands in one square centimetre of human skin.

300

The number of muscles the human body uses to balance itself while standing still.

450

The number of hairs in an average eyebrow.

500

The number of genes that women have more of than men, and because of this are protected from things like colour blindness and haemophilia.

500

Our kidneys filter about 500 gallons of blood each day.

500
The average human eye can distinguish about 500 different shades of grey.

540
The number of calories the average adult loses with every litre of sweat.

590
The number of miles of hair the average human will grow in a lifetime.

639
The number of muscles in the human body.

900
The human body contains enough carbon to make 900 pencils.

2,200
The human body contains enough phosphorous to make 2,200 match heads.

9,000
Of the ten thousand taste buds in the average human mouth, nine thousand are on the tongue (the other thousand are on the palate or in the cheeks).

10,000
Humans can distinguish up to 10,000 different smells.

60,000
The number of miles
of blood vessels in
the human body.

100,000
The number of hairs
on the average
person's head.

172,000
The skin of the armpits
can harbour up to
172,000 bacteria per
square centimetre.
Drier areas have only
about 4,000 bacteria
per square centimetre.

250,000
The number of sweat
glands in your feet (and
they sweat as much as 0.2
litres of moisture per day).

250,000
The number of round
trips of the body that
red blood cells make
before returning to the
bone marrow, where
they were born, to die.

700,000
The number of your
own skin flakes you
inhale every day.

2,000,000
The number of sweat
glands in the average
human body.

2,000,000
Our eyes are made
up of more than two
million working parts.

5,000,000
The number of red
blood cells in a tiny
droplet of blood.

15,000,000
The number of blood

cells produced and destroyed in the human body every second.

....

42,000,000
The number of times the average person's heart beats a year.

....

86,000,000
The number of bits of information per day that the human brain is capable of recording.

....

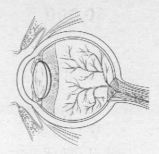

100,000,000
The number of light-sensitive cells in the retina.

....

100,000,000,000
The number of nerve cells in the human brain.

....

CELEBRITIES

0

The number of spleens Keanu Reeves, Burt Reynolds and Geoffrey Boycott have between them (they've all had their spleens removed).

0

The number of cups of coffee Steven Spielberg has ever had in his entire life.

1

The number of O levels achieved by Camilla, Duchess of Cornwall.

1

The number of shoelaces the film director David Lynch always leaves untied.

1

The number of castaways on *Desert Island Discs* who has been allowed to take a living person as a luxury – it was Dame Edna Everage who insisted on taking her bridesmaid, Madge Allsop. According to the presenter, Sue Lawley, 'I reminded her that the luxury had to be an inanimate object. She assured me that Madge was exactly that. So I allowed it.'

1

The number of kidneys Mel Gibson has (it's a fused horseshoe kidney).

1

The number of times
a week John Grisham
shaves (before church
on Sunday).

1

The number of times Elvis
Presley visited the UK (it
was in 1960 en route from
Germany to the US when
he stopped at Prestwick
Airport in Glasgow and
signed a few autographs).

2

The number of times
Chuck Berry was sent to
prison (for violating the
Mann Act – by taking a
girl across state borders
for 'immoral purposes'
– in 1962; for income
tax evasion in 1979).

2

The number of times
Paul Scofield refused
a knighthood.

2

The number of times
Kate Beckinsale won
the WH Smith Young
Writers Competition for
prose and poetry.

2

The number of webbed
toes on his left foot
Ashton Kutcher has.

2

The number of weeks
younger than Gary
Numan Gary Oldman is.

2

The number of butterflies Carol Vorderman's great-grandfather had named after him.

2

The number of months in jail Ozzy Osbourne served for burglary in 1966.

3

The number of times a week Jim Carrey's Labrador, Hazel, gets a professional massage (she lives in a $20,000 three-room dog house).

3

The number of times Cher's parents married – and divorced – each other.

3

The number of patents held by the American champion boxer Jack Johnson (the first patent was filed while Johnson was in jail at Leavenworth).

3

The number of years Samuel L. Jackson was Bill Cosby's stand-in on *The Cosby Show*.

3

The number of nipples Mark Wahlberg has (one was airbrushed out of the Calvin Klein underwear ad).

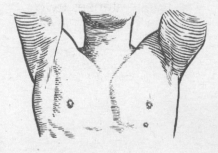

3

The number of days in jail Zsa Zsa Gabor spent in 1989 for slapping a cop.

3

The actress Salma Hayek was expelled from a Louisiana boarding school for setting alarm clocks back three hours.

3

The number of years in jail James Brown served, in two sentences (two years in 1988 for carrying a gun and assault, having previously served three years for theft when he was a teenager).

4

The number of schools Dizzee Rascal was expelled from in five years (in a fifth he was excluded from every class except music).

4

The number of hours Missy Elliott spends every day on her hair.

4

The number of Mickey Rooney's weddings that took place in Las Vegas (he was married eight times in total).

4

The number of years before Albert Einstein started speaking.

4

The number of counts on which Stephen King failed his army medical (flat feet, limited vision, high blood pressure and a punctured ear drum).

4

The number of years
for which the legendary
singer Edith Piaf was
blind (from the ages
of three to seven).

4

Clark Gable used to
shower more than
four times a day.

4

The number of years to
which Lord Jeffrey Archer
was sentenced in 2001
for perjury and perverting
the course of justice.

5

The number of sons
George Foreman
named George.

5

The number of octaves
Mariah Carey's vocal
range spans.

5

The number of gallons
of water Julio Iglesias
once had flown from
Miami to LA so he
could wash his hair.

6

The number of fingers
Gemma Arterton and
Sir Garfield Sobers
were born with on
each of their hands.

6

The number of TV screens
in Ice Cube's Navigator
sports utility vehicle.

6

The number of toes
the great darts player
Eric Bristow has on
one of his feet.

6

Kurt Cobain once bought six turtles and put them in a bathtub in the middle of his living room. When the smell got too bad, he and a friend drilled a hole in the middle of the floor as a drain.

7

The number of dwarfs Iggy Pop once requested to accompany him on stage.

8

The number of years before Stephen Hawking could read.

8

The number of dollars an hour Axl Rose used to earn for smoking cigarettes (for a science experiment at UCLA).

9

The number of days Sir Paul McCartney spent in a Japanese jail in 1980 for drug possession.

9

The number of siblings Susan Boyle has.

11

The number of siblings Dolly Parton has.

11

The number of times Russell Brand has been arrested.

12

The number of siblings Perry Como had (he was a seventh son of a seventh son).

13

The number of brothers and sisters Celine Dion had.

14

The number of schools Sylvester Stallone was expelled from in 11 years.

14

The number of siblings Charles Bronson had.

16

The singer/songwriter Dido once dropped 16 glasses of white wine and a tray on Stephen Fry when waitressing at Cafe Flo in London.

17

The number of days in prison to which Sophia Loren was sentenced in 1982 for tax evasion.

27

The number of children Evander Holyfield's father had.

27

Mel Gibson broke the school record for the most strappings in a week: 27.

45

The number of days Mark Wahlberg spent in jail at the age of 16 for his part in a robbery in which two Vietnamese men were beaten.

56

The number of curls child star Shirley Temple always had.

80

The percentage of Halle Berry's hearing in her left ear that was lost after an ex-lover beat her.

100

Oprah Winfrey once approached Aretha Franklin as she got out of a limo and convinced the singer that she was abandoned. Aretha gave her $100, which she used to stay in a hotel.

180

The IQ of the actor James Woods (way above MENSA entry level).

303

Katie Melua played the deepest underwater concert when she went 303 metres below sea level on Statoil's Troll A platform in the North Sea.

360

The degrees by which Daniel Radcliffe can rotate his arm.

400

Melissa Joan Hart can recite the mathematical expression 'pi' to 400 decimal places.

500

Penelope Cruz has more than 500 coat hangers in her collection.

1390

The number of minutes for which Garth Brooks signed autographs at the 1996 CMA Music Festival without taking a single break.

2,000

President Jimmy Carter (1977–81) developed the knack of reading at speed and was once tested and found to have a 95% comprehension rate reading at 2,000 words a minute.

6126
The name of Lindsay Lohan's fashion line.

25,000
The number of dollars raised by William Shatner for a housing charity when he auctioned a kidney stone.

50,000
The number of dollars W.C. Fields kept in Germany during World War II, 'in case the little bastard wins'.

100,000
The number of pounds Catherine Zeta-Jones's parents once won in a bingo game.

174517
Primo Levi's concentration camp number, which David Blaine has tattooed on his left forearm.

300,000
Dolly Parton insured each breast for $300,000.

600,000
Mariah Carey bought Marilyn Monroe's childhood piano for more than $600,000 at auction.

1,000,000
Keith Richards insured the third finger of his left hand for £1 million.

5,000,000
Bruce Springsteen insured his voice for $5 million.

1,000,000,000
Jennifer Lopez insured her body for $1 billion.

MUSIC

0

The number of encores Elvis Presley gave in his career.

0

The number of weeks the following songs spent in the British Top 20 – between them. *Angel Of The Morning* (P.P. Arnold), *Year Of The Cat* (Al Stewart), *What A Fool Believes* (The Doobie Brothers), *Someone Saved My Life Tonight* (Elton John), *We've Only Just Begun* (The Carpenters) *Stop Your Sobbing* (The Pretenders), *Last Train To Clarksville* (The Monkees), *I Don't Believe In Miracles* (Colin Blunstone) and *Wouldn't It Be Nice* (The Beach Boys).

1

The number of songs written by Larry Williams, a roadie for Neil Diamond, that were ever recorded (The Bellamy Brothers song *Let Your Love Flow*).

2

The number of fingers Django Reinhardt had missing on his right hand. He still became one of the greatest guitarists of all time.

2

The number of recorded tenors in the 20th century who were capable of singing E over high C (Enrico Caruso and Roy Orbison).

2

The highest number reached in the British charts by *Brown Sugar* by The Rolling Stones, *Golden Brown* by The Stranglers, *The Jean Genie* by David Bowie, *Delilah* by Tom Jones, *Lazy Sunday* by The Small Faces, *In The Air Tonight* by Phil Collins, *Oliver's Army* by Elvis Costello and the Attractions, *God Only Knows* by The Beach Boys, *When A Man Loves A Woman* by Percy Sledge and *In The Ghetto* by Elvis Presley.

3

The number of consecutive Christmas Number Ones the Spice Girls had – 1996: *2 Become 1*; 1997: *Too Much*; 1998: *Goodbye*.

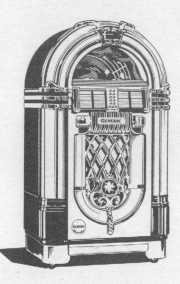

3

The number of Gershwin songs that contain the phrase 'Who could ask for anything more?' *I Got Rhythm, Nice Work if You Can Get It,* and *I'm About to Become a Mother.*

3

The number of acts that had British Number Ones with *Spirit In The Sky* (Norman Greenbaum (1970); Doctor And The Medics (1986); Gareth Gates featuring The Kumars (2003)).

3

The number of times 'Do They Know It's Christmas?' reached number one in the UK (1984, 1989 and 2004).

4

At the 2010 Grammy Awards, Taylor Swift won four Grammys – one more than Elvis did in his entire career.

5

Sir Cliff Richard had British Number One singles in five different decades (the 50s, 60s, 70s, 80s and 90s).

5

The Beatles held the Top Five spots on the 4 April, 1964 US *Billboard* singles chart. To date, they're the only band to have accomplished that feat.

6

Karen Carpenter's doorbell chimed the first six notes of *We've Only Just Begun.*

6

The length in hours of the longest opera performed at the Royal Opera House (Richard Wagner's *Götterdämmerung*).

7

The number of countries that participated in the very first Eurovision Song Contest. It was held in Lugano, Switzerland, on 24 May 1956 and each country sang two songs (the only time this has happened). The host nation won.

10

The number of minutes it took George Gershwin and Irving Caesar to write the song *Swanee* – which sold over two million copies in 1919 after being recorded by Al Jolson.

19

The number of Hungarian rhapsodies written by Franz Liszt.

20

The number of years ago today that Sergeant Pepper taught the band to play (in *Sgt. Pepper's Lonely Hearts Club Band*).

24

The number of consecutive Top 5 singles in the US achieved by Elvis (his run ended with *Return To Sender* in 1962).

29

The number of the track from where the *Chattanooga Choo-Choo* could be boarded in the 1941 song.

30

The number of artistes who turned down the

song *Bye, Bye Love* before The Everly Brothers recorded it. It became their first big hit.

47
The number of strings on a harp.

49
The number in the UK charts reached by the group Jet Bronx And The Forbidden – featuring Loyd Grossman as the singer and guitarist – with *Ain't Doin' Nothin'* in 1977.

50
The number of ways to leave your lover, according to Paul Simon.

60
The number of vocalists The Drifters have had.

76
The number of trombones that led the big parade in *The Music Man*.

88
The number of keys on a piano.

104
The number of symphonies composed by Haydn.

120
The number of hours it would take to play the complete works of Beethoven.

165
The (record) number of
curtain calls received
by Luciano Pavarotti,
in 1988 in Berlin.

175
The number of hours
of music composed
by JS Bach.

202
The number of hours
of music composed
by Mozart.

273
The number of seconds
of silence in John Cage's
composition *4'33*.

447
The length in seconds
of the shortest opera
ever published (*The
Deliverance of Theseus*
by Darius Milhaud).

461
On 14 February, 1977,
singer/songwriter Janis Ian
received 461 Valentine's
Day cards after the success
of her song *At Seventeen*,
which contained the
lyrics 'The valentines
I never knew'.

4,000
The number of holes in
Blackburn, Lancashire in
A Day in the Life.

ART & LITERATURE

1

The number of Shakespeare plays that don't have songs in them (*The Comedy of Errors*).

1

The number of Nobel Literature laureates to have made it into Wisden (Samuel Beckett played in a first-class fixture for Dublin University against Northamptonshire in 1926, scoring 18 and 12).

1

The number of full-length novels Kenneth Grahame wrote (*The Wind In The Willows*).

1

The number of bottles of ink Victor Hugo used to write *The Hunchback of Notre Dame*.

1

The number of novels Emily Brontë wrote (*Wuthering Heights*).

1

The number of novels Harper Lee wrote (*To Kill A Mockingbird*).

1

The number of novels Margaret Mitchell wrote (*Gone With The Wind*).

1

The number of dogs in Shakespeare plays (Crab, in *The Two Gentlemen of Verona*).

1

The number of his sculptures Michelangelo signed (*The Pieta*, completed in 1500).

1

The number of paintings sold by Vincent van Gogh in his lifetime.

2

The number of people who 'beat' Graham Greene in a competition in *The Spectator* to parody his style (he himself entered and came third).

3

X-ray technology has shown there are three different versions of the *Mona Lisa* under the visible one.

4

The number of Shakespeare plays with ghosts in them (*Julius Caesar, Richard III, Hamlet* and *Macbeth*).

4

The number of poems of the 900 that Emily Dickinson wrote that were published in her lifetime.

4

The number of 'good' legs in George Orwell's *Animal Farm*.

5

The number of desks that Ted Hughes had in his study.

6

When Leonardo da Vinci's *Mona Lisa* was stolen from the Louvre in 1912, six replicas were sold as the original in the three years before the original was recovered.

7

The number of times Leo Tolstoy's wife copied his manuscript of *War and Peace* by hand.

9

The number of circles of hell in Dante's *The Divine Comedy*.

10

The number of legs shared by the four women in the Edgar Degas painting *Young Spartans Exercising*.

13

The number that the clocks were striking in the opening line of George Orwell's *1984*.

13

'On top of everything, the cancer wing was number 13' (the first line of *Cancer Ward* by Alexandr Solzhenitsyn).

14
The number of kisses in the collected novels written by Jane Austen. Three between a man and a woman, four between females, two from women to children, four on the hand and one by a man to a lock of severed hair.

15
The number of men on a dead men's chest in Robert Louis Stevenson's *Treasure Island*.

17
The number of syllables in a haiku.

18
Joseph Heller's original title for *Catch-22* was *Catch-18*.

20
The number of years Rip Van Winkle slept.

20
The title for Peter Benchley's best-selling novel *Jaws* wasn't agreed until 20 minutes before the book went into production.

35
The length in seconds of the shortest stage play (Samuel Beckett's *Breath*).

37
The number of plays written by Shakespeare.

40
The number of thieves encountered by Ali Baba.

47

The number of days in 1961 that the Museum of Modern Art in New York City hung Matisse's *Le Bateau* upside-down before an art student noticed the error.

50

The number of copies of *Moby Dick* sold during its writer Herman Melville's lifetime.

57

The number of years older than Juliet the legendary actress Sarah Bernhardt was when she played her at the age of 70.

59

The number of Enid Blyton stories published in 1959.

90

The minimum number of self-portraits by Rembrandt.

117

'There were 117 psychoanalysts on the Pan Am flight to Vienna and I'd been treated by at least six of them' (the opening line of *Fear of Flying* by Erica Jong).

142

The number of staircases at Hogwarts School in the Harry Potter books.

156

The number of fairy tales by Hans Christian Andersen.

237

The number of lines Iago (1,097 lines) has more than Othello (860 lines) in *Othello*.

1,614

The number of lines Falstaff had in *Henry IV, Parts 1 & 2* and *The Merry Wives of Windsor* – the most of any Shakespeare character. Hamlet had 1,422 lines in *Hamlet* – making Hamlet the longest role in any *single* Shakespeare play.

6,000

Alfred, Lord Tennyson wrote a 6,000-word epic poem when he was just 12 years old.

8,000

The number of different words John Milton used in *Paradise Lost*.

13,955

The number of words in the longest sentence in literature. It is in *The Rotters' Club* by Jonathan Coe.

17, 677

The number of different words used by Shakespeare.

MOVIES & TV

1

The number of years *younger* than Cary Grant Jessie Royce Landis was when she played his mother in the Hitchcock classic *North By Northwest.*

1

The number of members of the cast of *Dad's Army* to have served in the Home Guard (John Laurie).

1

The number of recipients of Golden Globe Awards for 'Most Glamorous Actress'. Zsa Zsa Gabor won the award in 1958 but the category was cancelled afterwards.

1

The number of the cast (and crew) of the *Lord of the Rings* movies to have met J.R.R. Tolkien (Christopher Lee).

1

The number of soundtracks to out-gross the movie itself – *Superfly* (1972).

1

In *The Bridge on the River Kwai* (1957), Alec Guinness – who won an Oscar for his role – had his name spelt with just one 'n' in the final credits.

2

The number of animated Disney full-length features that have both parents present and alive throughout the movie (*101 Dalmatians* and *Peter Pan*).

2

The number of actors who turned down the role of Captain Mainwaring in *Dad's Army* (Arthur Lowe was only cast after Thorley Walters and Jon Pertwee turned it down).

2

The number of times Hitler watched Chaplin's *The Great Dictator*. He'd banned it but he was curious to see the film himself and so he had a print of the film smuggled into Germany from Portugal and watched it not once, but twice.

3

The number of patients to appear in each episode of *Dr Kildare*.

3

The number of fingers on each character's hand in animated movies. A film animator explains: 'In the early days of animation when everything had to be done by hand, animators would have to draw thousands of pictures and it was easier to do characters with three fingers instead of four. Also, you have to remember that animation was anyway a lot more basic in the beginning. They weren't trying to achieve precision; just a general sense of a character.'

5

The number of times Francis Ford Coppola was fired while directing *The Godfather*.

9

The number of kilograms
Natalie Portman lost
to play Nina Sayers in
Black Swan (2010).

9

In *The Simpsons,* Bart's
hair has precisely
nine spikes.

9

The number of kilograms
Beyoncé Knowles lost
to play Deena Jones in
Dreamgirls (2006).

13.6

The number of kilograms
Edward Norton put on
to play Derek Vinyard in
American History X (1998).

13.6

The number of kilograms
Matt Damon put on to
play Mark Whitacre in *The
Informant* (2009).

13.6

The number of kilograms
Charlize Theron put on
to play Aileen Wuornos
in *Monster* (2003).

13.6

The number of kilograms
Taylor Lautner put on to
reprise the role of Jacob
Black in *The Twilight
Saga: New Moon* (2009).

15.87

The number of kilograms
George Clooney put
on to play Robert Baer
in *Syriana* (2005).

19

The number of kilograms Tom Hardy put on to play Charles Bronson/Michael Peterson in *Bronson* (2008).

22

The Paramount logo contains 22 stars.

22.6

The number of kilograms Tom Hanks lost to play Chuck Noland in *Castaway* (2000).

22.6

The number of kilograms Jennifer Hudson lost to play Winnie Mandela in *Winnie* (2011).

24

The number of kilograms 50 Cent lost to play Deon in *All Things Fall Apart* (2011).

25

The number of pounds Peter Sellers was paid to provide the voices for the first PG Tips chimps' TV commercial.

27

The number of kilograms Robert De Niro put on to play Jake La Motta in *Raging Bull* (1980).

28

The number of kilograms Christian Bale lost to play Trevor Reznik in *The Machinist* (2004), and then put double that back on to play Batman/Bruce Wayne in *Batman Begins* (2004).

28.5

The number of kilograms Russell Crowe put on to play Ed Hoffman in *Body of Lies* (2008).

31.75

The number of kilograms Vincent D'Onofrio put on to play Private Leonard 'Gomer Pyle' Lawrence in *Full Metal Jacket* (1987).

45

The length in centimetres of the model of the ape in the 1933 film *King Kong*.

47

The number of piglets used in the filming of *Babe*.

50

The number of hard-boiled eggs eaten by Paul Newman's character in *Cool Hand Luke*.

67

The number of years the actor David Manners lived for after co-starring with Bela Lugosi in the 1931 film *Dracula* without ever watching it.

73

The number of performers nominated for an Oscar in their (screen) debut role.

79

The number of original *Star Trek* episodes. In the pilot, *The Cage*, the Captain (Captain Pike) was played by Jeffrey Hunter. When the TV series got the go-ahead, Hunter declined to be in it (rumour has it that his wife advised him against it) and William Shatner (Captain Kirk) was picked instead.

80

The number of cigarettes a day Vivien Leigh smoked during the filming of *Gone With The Wind*. Clark Gable smoked 'just' sixty a day.

185

The number of seconds the longest movie kiss lasted. The film was *You're In The Army Now* (1941) and the kissers were Jane Wyman (Ronald Reagan's first wife) and Regis Toomey.

191

The number of kisses John Barrymore performed with different women in the 1926 film *Don Juan*. This

was before the 1930 Hays Code, which banned 'excessive and lustful kissing.'

262

The number of times Eddie Murphy swore in the film *Harlem Nights*.

400

The number of actresses who tested for the role of Scarlett in *Gone With The Wind*, out of 1,400 who were interviewed. They included Lana Turner, Bette Davis, Norma Shearer, Lucille Ball, Tallulah Bankhead, Claudette Colbert, Katharine Hepburn, Joan Crawford, Loretta Young, Jean Harlow and Carole Lombard. However, only Paulette Goddard and Vivien Leigh were filmed in colour for their auditions.

742

The house number in Evergreen Terrace, Springfield, where the

Simpsons live. No one knows where Springfield, the Simpsons' home town, is and the writers often tease the audience. For example, in one episode, Marge phones an egg-painting company and, when asked for her address, says, '742 Evergreen Terrace, Springfield, Ohi-', then stops herself and says, 'Oh hiya Maude!' to Maude Flanders, who suddenly appears in her kitchen.

2211
Dirty Harry's badge number.

4,077
The number of the Mobile Army Service Hospital in *M*A*S*H*.

8,000
The (largest) number of horses assembled for King Vidor's epic film *War and Peace* (1956).

20,000
The highest-paid animal actors are bears, which can earn $20,000 a day.

24601
Sideshow Bob's prison number in *The Simpsons* – the same as Jean Valjean's prison number in *Les Miserables*.

233,500
The amount in dollars the Rosebud sled from *Citizen Kane* fetched at auction.

398,500

The amount in dollars
The Maltese Falcon icon
from the film of the same
name fetched at auction.

666,000

The amount in dollars
Dorothy's Ruby Slippers
from *The Wizard of Oz*
fetched at auction.

193,000,000

The amount in dollars
more it cost to make a
film about the *Titanic* than
it did to actually build it.

30,150,000

The number of people
who tuned in to watch
the Christmas Day 1986
edition of *EastEnders*.

248,000,000

The number of dollars
The Blair Witch Project
grossed having only cost
$22,000 to make – thus
making it the film with
the highest box-office
sales-to-production
cost ratio of all time.

OSCARS

0

The number of Oscars won by Gene Kelly, Steve McQueen, Cary Grant, Glenn Ford, James Mason, Anthony Quayle, Montgomery Clift, Kirk Douglas, Greta Garbo, Barbara Stanwyck, Lana Turner, Judy Garland, Lee Remick, Natalie Wood, Rita Hayworth and Gloria Swanson *between* them.

0

The number of Oscars won by these classic films *between* them:

Bad Day At Black Rock (1955)

•

The Sixth Sense (1999)

•

Taxi Driver (1976)

•

The Elephant Man (1980)

•

Being John Malkovich (1999)

•

Brief Encounter (1946)

•

The Talented Mr Ripley (1999)

•

To Be Or Not To Be (1942)

•

Deliverance (1972)

•

The Magnificent Ambersons (1942)

•

I Am A Fugitive From A Chain Gang (1932)

•

The Shawshank Redemption (1994)

The Great Escape (1963)
•
American Graffiti (1973)
•
A Clockwork Orange (1971)
•
Top Hat (1935)

Dr Strangelove (Or How I Learned To Stop Worrying And Love The Bomb) (1964)
•
Psycho (1960)
•
12 Angry Men (1957)
•
The Green Mile (1999)
•
Vertigo (1958)
•
The Maltese Falcon (1941)
•
The Player (1992)

Rebel Without A Cause (1955)
•
The Caine Mutiny (1954)
•
The Great Dictator (1940)
•
The China Syndrome (1979)
•
When Harry Met Sally (1989)

Cat On A Hot Tin Roof (1958)
•
Rear Window (1954)
•
A Star Is Born (1954)

Alfie (1966)
•
North By Northwest (1959)
•
A Few Good Men (1992)
•
Singin' In The Rain (1952)
•
Amelie (2001)

....

0

The number of Oscar nominations won by these classic films *between* them:

Saturday Night & Sunday Morning (1960)
•
The Wild One (1953)
•

This Is Spinal Tap (1984)

•

Kind Hearts And Coronets
(1948)

•

The Cincinnati Kid (1965)

•

King Kong (1933)

•

O Lucky Man (1972)

•

My Darling Clementine
(1946)

•

Mean Streets (1973)

•

A Night At The Opera
(1934)

•

The Big Sleep (1946)

•

The Misfits (1961)

•

The 39 Steps (1935)

•

Sweet Smell Of Success
(1957)

•

Oliver Twist (1948)

•

His Girl Friday (1940)

•

Reservoir Dogs (1992)

•

The Night Of The Hunter
(1955)

•

The Searchers (1956)

•

Dirty Harry (1971)

•

*The Postman Always Rings
Twice* (1946)

•

Bringing Up Baby (1938)

•

Paths of Glory (1957)

•

Modern Times (1936)

•

The Lady Vanishes (1939)

•

*Once Upon A Time
In America* (1984)

1

The number of Nobel
prize-winners to win
Oscars (George Bernard
Shaw for *Pygmalion*).

1

The number of British footballers to win Oscars. Neil Paterson played for Dundee United in the 1940s while also working as a freelance writer. In 1959, he won the Oscar for his screenplay for *Room At The Top*.

1

The number of people to open the Oscar envelope to find their own name on the card: Irving Berlin (Best Song, *White Christmas* 1942).

1

The number of people to be nominated for producer, director, writer and actor on the same film – twice. Warren Beatty for *Heaven Can Wait* (1978) and *Reds* (1981).

1

The number of roles to garner two Oscars (Vito Corleone). Marlon Brando won the Best Actor Oscar for his role in *The Godfather* (1972) while Robert De Niro won the Best Supporting Actor Oscar in the same role for *The Godfather Part II* (1974).

1

The number of men named Oscar to win an Oscar: Oscar Hammerstein II (Best Song: 1941; 1945).

1

The number of mothers and daughters to be nominated for Oscars in the same year: Diane Ladd and her daughter Laura Dern for *Rambling Rose* (1991).

2

The number of families in which three generations have won Oscars: The Hustons: Walter for *The Treasure Of The Sierra Madre* (1948), John for *The Treasure Of The Sierra Madre* and Anjelica for *Prizzi's Honor* (1985). Walter and John Huston are also the only father and son to win acting Oscars for the same film. The Coppolas: Carmine for *The Godfather, Part II* (1974), his son Francis for *The Godfather, Part II* (1974) and Sofia for *Lost In Translation* (2004). In addition, Francis's nephew (Sofia's cousin) Nicolas Cage won the Best Actor Oscar for *Leaving Las Vegas* (1995).

2

The number of non-professional actors to win acting Oscars: Harold Russell (*The Best Years Of Our Lives*, 1946) and Dr Haing S. Ngor (*The Killing Fields*, 1984).

2

The number of Oscar categories for which women have never won prizes: Best Sound and Best Cinematography.

2

The number of actresses to win consecutive Best Actress Oscars: Luise Rainer (1936 and 1937) and Katharine Hepburn (1967 and 1968).

3

The number of films in which all the members of the cast have been nominated for Oscars: *Who's Afraid Of Virginia Woolf?* (four cast members, 1966), *Sleuth* (two cast members, 1972) and *Give 'em Hell Harry* (one cast member, 1975).

3

The number of actors to win consecutive Oscars: Spencer Tracy (1937 and 1938); Jason Robards (1976 and 1977); and Tom Hanks (1993 and 1994). Tracy and Hanks won Best Actor Oscars; Robards won Best Supporting Actor Oscars.

3

The number of films to win Oscars for Best Picture, Best Director, Best Actor, Best Actress and Best Screenplay: *It Happened One Night* (1934); *One Flew Over the Cuckoo's Nest* (1975); and *The Silence Of The Lambs* (1991).

4

The number of films to be nominated for nine or more Oscars and to win Oscars in every category for which they were nominated: *Lord of the Rings: The Return of the King* (11/11 Oscars in 2004); *The Last Emperor* (9/9 in 1988); and *Gigi* (9/9 in 1959).

4

The number of actresses to win Best Actress Oscars for their very first on-screen appearances: Shirley Booth (*Come Back Little Sheba*, 1953); Julie Andrews (*Mary Poppins*, 1965); Barbra Streisand (*Funny Girl*, 1969); and Marlee Matlin (*Children Of A Lesser God* 1987).

4

The number of people to win an acting Oscar and to have a number-one album in the US (Frank Sinatra, Bing Crosby, Barbra Streisand and Jamie Foxx).

4

The number of years running Marlon Brando (1951–4) and Al Pacino (1972–5) were nominated for Oscars.

....

5

The number of years running Bette Davis (1938–42) and Greer Garson (1941–5) were nominated for Oscars.

....

6

The number of years double-Oscar winners are statistically likely to outlive other actors by.

....

7

The number of films to win Best Actor and Best Actress Oscars: *It Happened One Night* (1934); *One Flew Over the Cuckoo's Nest* (1975); *Network* (1976); *Coming Home* (1978); *On Golden Pond* (1981); *The Silence Of The Lambs* (1991); and *As Good As It Gets* (1997).

8

The number of people to win the Tony and then the Oscar for the same role:

José Ferrer for *Cyrano de Bergerac* (Tony: 1947/ Oscar: 1950)

•

Shirley Booth for *Come Back, Little Sheba* (1950/1953)

•

Yul Brynner for *The King and I* (1952/1956)

•

Rex Harrison for *My Fair Lady* (1957/1964)

•

Anne Bancroft for *The Miracle Worker* (1960/1962)

•

Paul Scofield for *A Man for All Seasons* (1962/1966)

•

Jack Albertson for *The Subject Was Roses* (1965/1968)

Joel Grey for *Cabaret*
(1967/1973)

•

NB. Lila Kedrova did it the
other way around. She
won a 1964 Oscar for
Zorba the Greek, and 20
years later won a Tony for
the same role in *Zorba*.

11

The number of Oscar
nominations *The Turning
Point* (1977) and *The
Color Purple* (1985)
won without winning a
single Oscar (a dubious
record they both share).

10

The number of people
who have been
nominated for two acting
Oscars in the same
year – the most recent
being: Holly Hunter and
Emma Thompson (1993);
Julianne Moore (2002);
and Jamie Foxx (2004).

32

The number of Oscars
won by Walt Disney
(more than anyone
else in history).

FOOD & DRINK

0.028

On average, a drop of Heinz tomato ketchup leaves the bottle at a speed of 0.028 miles per hour.

1

The number of fruits that have their seeds on the outside (the strawberry).

3.2

The number of kilograms of tea the average Irish adult drinks annually (more than any other nation).

4

A can of Spam is opened every four seconds.

4.5

The number of cups of coffee the average citizen of Finland drinks daily (more than any other nation).

5

The number of years Shredded Wheat was available before Kellogg's Corn Flakes were first introduced.

6

Every year enough potatoes are grown worldwide to cover a four-lane motorway circling the world six times.

8

The number of cows eaten by the average Briton over a lifetime.

8.5

The number of kilograms of butter the average French person consumes annually (more than any other nation).

9

The number of yellow kernels for every one white kernel in a bag of popcorn.

10

The number of kilograms of milk it takes to make one kilogram of cheese.

11

The total number of cases of Chablis produced by the whole region in 1957 because of the terrible springtime frost.

20

The percentage of the world's peanut crop used by chocolate manufacturers.

20

The number of bottles in a Nebuchadnezzar.

25

The number of bottles of Coca-Cola sold in its first year (1886).

26

Of the 1,000+ chemicals in a cup of coffee, only 26 have been tested (half of these caused cancer in rats).

27
The number of kilograms of pasta the average Italian consumes annually (more than any other nation).

30
The number of blends of tea in the average teabag.

32
The percentage of the world's rice eaten by the Chinese.

35
During a lifetime, the average person eats 35 tons of food.

40
The percentage of the world's almond crop used by chocolate manufacturers.

50
The number of cups of coffee a day drunk by the French philosopher, Voltaire.

50
The number of glasses of chocolate a day drunk by Montezuma, an Aztec Emperor of Mexico.

59.29
The number of seconds it took Simon Sang Sung of Singapore to turn a single piece of dough into 8,192 noodles.

60
Beethoven was so particular about his coffee that he always counted out 60 beans for each cup.

62
The weight in grams of the largest gooseberry ever grown.

66.6
The percentage of the world's aubergines grown in New Jersey.

75
The percentage of the world's supply of maple syrup that comes from Canada.

90
The percentage of the vitamin C in Brussels sprouts that is lost in cooking.

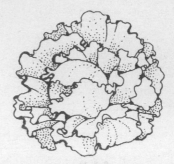

91
The percentage of water in cabbage.

92
The percentage of water in watermelons.

96
The percentage of water in a cucumber.

123
The number of snails the average French citizen eats a year.

143

The number of kilograms of potatoes the average Irish person eats annually (more than any other nation).

165

The number of litres of milk the average Irish person drinks annually (more than any other nation).

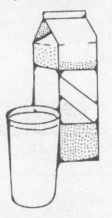

178

The average number of sesame seeds on a Big Mac bun.

300

The number of distinctly different types of honey.

300

France has more than 300 different kinds of cheese (and the French eat more cheese than any other nation: an average of 23 kg per person per year).

400

The number of quarter-pound hamburgers that can be made out of one cow.

548

The number of peanuts in a 340g jar of peanut butter.

800

The number of kernels an average ear of corn has, arranged in sixteen rows.

800

There are over 800 different varieties of beer brewed in Belgium.

800
The amount of garlic in grams the average French person eats a year.

2,000
There are more than 2,000 different types of cheese in the world.

4,000
The number of crocuses required to produce a single ounce of saffron.

9,000
The number of Indian restaurants in the UK.

10,000
It costs 10,000 times more to produce a litre of bottled water than a litre of tap water.

10,000
The number of chocolate bars the average person will eat in a lifetime.

13,345
The number of eggs the average person will eat in a lifetime.

16,000
A human being drinks 16,000 tons of water in a lifetime.

20,000
The number of living organisms in a glass of water.

27,000
During your lifetime, you'll eat about 27,000 kilos of food.

40,000
There are more than 40,000 different varieties of rice.

70,000
The average person drinks 70,000 cups of coffee in a lifetime.

200,000
There are sweeteners that are 200,000 times sweeter than sugar.

3,000,000,000
The water we drink is three billion years old.

400,000,000,000
Coffee is the most popular beverage worldwide with over 400 billion cups consumed each year.

MONEY

6

The number of Americans who, in 1911, each paid $300,000 for the *Mona Lisa*. All the paintings had been painted by the Frenchman, Yves Chaudron.

10

The percentage of gift cards or vouchers that are not redeemed.

15

In 1849, Walter Hunt owed a friend $15. The friend offered to cancel the debt if Hunt could fashion a working item from a piece of wire. After a while, Walter had created the safety pin – and made his friend into a millionaire.

29

The number of cash registers in the world's largest McDonald's (in Beijing, China).

30

The number of days it takes for one penny doubled daily to become over £5 million.

35

The number of dollars Caroline Davidson was paid in 1971 to create the Nike swoosh logo.

51

The number of razor sets sold by Gillette in the first year of production.

130

The number of dollars Joe Shuster and Jerry Siegel received in 1938 when they sold all rights to the comic-strip character Superman to their publishers.

172

The number some say can be found in the bushes at the base of the Lincoln Memorial pictured on the back of the US $5 bill.

250

The average American bank teller loses about $250 every year.

378.79

The amount in dollars per person Bob Fosse, the choreographer and film director, left to each of 66 people to 'go out and have dinner on me'. They included Liza Minnelli, Janet Leigh, Elia Kazan, Dustin Hoffman, Melanie Griffith, Neil Simon, Ben Gazzara, Jessica Lange and Roy Scheider.

500

The amount in pounds per person Ian Fleming, the author of the James Bond novels, left to each of four friends with the instruction that they should 'spend the same within twelve months of receipt on some extravagance'.

1,513
The number of dollars a 200-year-old piece of Tibetan cheese fetched at auction.

2,300
The number of dollars received by Dr John Pemberton, the man who invented Coca-Cola in 1896, when he sold his formula for the drink in 1897.

6,600
The number of dollars fetched by Lee Harvey Oswald's cadaver tag at auction.

100,000
The number of dollars spent by American researchers on making the discovery that three out of four people hang their toilet rolls so that the paper is pulled down to be torn off rather than up.

445,500
The number of dollars John Lennon's handwritten lyric to the Beatles' song *Nowhere Man* fetched at auction.

497,500
The number of dollars Eric Clapton's 1956 Fender Stratocaster (as used on *Layla*) fetched at auction.

1,300,000
The most money in US dollars ever paid for a cow in an auction.

6,666,666
The phone number 666 6666 fetched $2.75 million at an auction in Qatar (the number six is considered lucky in Qatar).

2,000,000,000

The world spends more than $2 billion a day on weaponry.

41,900,000, 000,000,000

The percentage inflation in Hungary in July 1946.

1,000,000,000, 000,000,000,000

The largest banknote ever issued was the 1946 Hungarian 100 quintillion pengo note. That's a one with twenty-one zeros.

NUMBERS

1

The only number with its letters in reverse alphabetical order.

6

A perfect number. The numbers that go into six – 1, 2 and 3 – add up to six.

11½

The number of days it would take – counting at one digit per second – to count to one million.

13

The number in a baker's dozen. This is why in the 13th century, there was a law called the Assize of Bread and Ale in which it was decreed that bakers who short-changed their customers faced serious punishment – like having an arm chopped off. So, rather than risk such a fate, bakers would routinely hand over thirteen items when asked for a dozen – just in case they'd miscounted or one of the items was defective in some way.

13

The smallest number with eight letters.

14

The calendar repeats itself every 14 years.

15

The smallest number with seven letters.

23

The number of people who need to be in the same room to make it probable that two of them will share the same birthday. But although there only have to be 23 people present to make it probable that any two share the same birthday, if you specify the date of that birthday they share – say 11 February – then there have to be 253 people present to make it probable.

37

The numbers 111, 222, 333, 444, 555, 666, 777, 888 and 999 are all multiples of 37.

40

The only number with its letters in alphabetical order.

52

The number of letters in Acetwothreefourfivesixseveneightninetenjackqueenking – the same as the number of cards in the deck.

153

A number which is the 'sum of the cubes of its digits' (i.e. 153 = 1+125+27).

293

The number of different ways there are to make change for a dollar.

364

The total number of gifts that 'my true love gave to me' in the carol 'The Twelve Days of Christmas'.

600

A 'centillion' is the number 1 followed by 600 zeros (300 in the US).

1,000

This number contains the letter A (none of the words from one to nine hundred and ninety-nine does).

1,089

If you multiply 1,089 by 9 you get 9,801. It's reversed itself. This also works with 10,989 or 1,099,989 and so on.

1666

The Roman numerals for this are MDCLXVI (1000+500+100+50+10+5+1.) and the year is famous as being the one time in history when the date is written with all of the Roman numerals from the highest value to the lowest value.

2,520

The number 2,520 can be divided precisely by 1, 2, 3, 4, 5, 6, 7, 8, 9, and 10.

5,050

The total when you add up the numbers 1 to 100 consecutively (1 + 2 + 3 + 4 + 5, etc.).

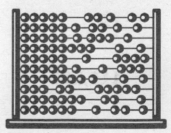

10,000

The ancient Greeks had no way to describe any

numbers more than 10,000 and so simply called them a myriad.

....

21,978

If you multiply 21978 by 4, you get 87912 – which is the digits 21978 in reverse order.

....

37,037

Multiply 37,037 by any single number (1–9), then multiply that number by 3. Every digit in the answer will be the same as that first single number.

....

115,200

The number of possible permutations of six couples at a twelve-person table so that men and women alternate and are also not sitting next to their partners.

....

142,857

When 142,857 is multiplied by any number from one to six, the result is a number containing the same digits *in the same order* – but starting in a different place.

$1 \times 142,857 = 142,857$

•

$2 \times 142,857 = 285,714$

•

$3 \times 142,857 = 428,571$

•

$4 \times 142,857 = 571,428$

•

$5 \times 142,857 = 714,285$

•

$6 \times 142,857 = 857,142$

•

When you multiply 142,857 by 7: it equals 999,999

•

When 1 is divided by 7 it comes to 0.142857,142857,142857

•

If you multiply 142,857 by 8, you get 1142856. If you take off the first digit (1) and add it to the remaining six digits, see what happens:
$1 + 142856 = 142857$

•

Similarly, if you multiply 142,857 by – say – 17, you get 2,428,569. Now take off the first digit (2) and add it to the remaining six digits and see what happens: 2 + 428,569 = 428,571 – which is, of course, the original number in stage three of its cycle.

•

You can do this with every number except multiples of seven (although these also produce interesting results).

•

If you take the number 142,857 and split it into two – 142 and 857 – and then add those two numbers, you get 999.

999,999
The 762nd to 767th digits of pi are 999999.

1,274,953,680
The number that uses all the digits 0–9 that can also be divided by any number from 1 to 16 without leaving a remainder.

31,557,600
The number of seconds in a year.

RELIGION & THE BIBLE

0
The number of cats or rats mentioned in the Bible.

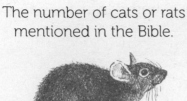

0
The number of times that angels' wings are mentioned in the Bible.

1
The number of times the word 'girl' appears in the Bible.

1
The number of Apostles who died a natural death (St John).

1
The number of books in the Bible that don't mention the name of God (the book of Esther).

2
The number of times pigs are mentioned in the Bible.

2
The number of times cheese is mentioned in the Bible.

2
The number of different nuts mentioned in the Bible (almonds and pistachios).

3

The number of angels
mentioned by name
in the Bible (Gabriel,
Michael and Lucifer).

3

The number of days
Jonah spent in the
belly of a whale.

3

The smallest church in
the world is in Kentucky.
There's room inside
for just three people.

4

The number of Noble
Truths of Buddhism.

6

The number of women
named Mary in the Bible.

7

The number of suicides
recorded in the Bible.

7

The number of archangels
(Michael, Gabriel,
Raphael, Uriel, Chamuel,
Jophiel and Zadkiel).

9

The number of choirs
of angels (from highest
to lowest – they are:
seraphim, cherubim,
thrones, dominions,
virtues, powers,
principalities, archangels
and angels).

18

The number of letters
in Mahershalalhashbaz
– the longest word in
the Bible (Isaiah 18:1).

35

The number of days on which Easter may fall. Originally, Easter was celebrated on the same day as the Jewish Passover. Since the Jewish calendar is lunar, Passover can fall on any day of the week. The Church wanted to ensure that Easter would at least fall on a Sunday. In the 8th century, the Church officially adopted the formula: Easter Day is the first Sunday subsequent to the first full moon after the vernal equinox (21 March – as early as 22 March and as late as 25 April, if the full moon after the equinox falls on a Sunday, Easter follows a week later).

49

The number of different kinds of food mentioned in the Bible.

62

The number of times camels are mentioned in the Bible.

66

The number of books in the St James Bible.

68

The number of words in The Lord's Prayer.

90

The percentage of British adults who owned a Bible sixty years ago (today, that figure is closer to 50%).

118

The number of the Psalm containing the middle two verses of the Bible.

164

The number of times horses are mentioned in the Bible.

176

The number of times lions are mentioned in the Bible.

188

The number of times lambs are mentioned in the Bible.

250

The number of calfs required to provide the skin for the Winchester Bible, the largest surviving 12th-century English Bible.

260

The number of chapters in the New Testament (containing 181,253 words).

666

The number of the Beast (according to the Book of Revelations 13, 18: 'if anyone has insight, let him calculate the number of the beast for it is the man's number. His number is 666').

929

The number of chapters in the Old Testament (containing 592,439 words).

1,700

The number of references to gems and precious stones in the Bible.

2,200

The number of languages and dialects into which the Bible has been translated.

46,277

The number of times the word 'and' appears in the Bible.

HISTORY

1

The number of European countries occupied by the Germans that ended World War II with a larger Jewish population than before the start of the War: Albania. Only one Jewish family was deported and killed during the Nazi occupation of Albania. Not only did the Albanians protect their own Jews, but they provided refuge for Jews from neighbouring countries.

1

Medieval Welsh mercenary bowmen only wore one shoe at a time.

1

The number of months in recorded history not to have a full moon (February 1865).

1

The number of times that Canada has declared war on another country (it was World War II).

1

Pope Stephen II served for just one day in the year 752.

1

The number of dogs registered as POWs in World War II (a Pointer named Judy).

1

The number of testicles that Adolf Hitler had. The evidence for this is as follows: in World War I, during the Battle of the Somme, Hitler was shot – either in the thigh or the groin. Later, his company commander claimed that he'd discovered that Hitler was missing a testicle during a routine medical inspection. Hitler killed himself in 1945 and the Russians carried out an autopsy. The report of this autopsy was published in 1968 and revealed that Hitler's 'left testicle could not be found either in the scrotum or on the spermatic cord inside the inguinal canal or in the small pelvis...'. This autopsy was corroborated in 1972 by a team from the University of California.

1

From 1925 to 1945, Hitler held the official title of SS Member #1. The man who was Member #2 wasn't Heinrich Himmler but Emil Maurice, Hitler's personal bodyguard/chauffeur and the man who was credited with founding the SS. Maurice, incredibly, was half-Jewish and, when it came to light in 1935, he was thrown out of the SS. However, he was allowed to retain all his privileges.

2

In 1531, the Bishop of Rochester's chef was sentenced to death for poisoning members of the bishop's household. He was taken to the town square where he was put into a pot of cold water hanging over a fire. It took two hours for the water to boil and kill him.

2

The number of times the River Nile has frozen over: once in the ninth century, and then again in the eleventh century.

2

After Attila the Hun murdered his wife's brothers, she served him the hearts of his two sons for dinner.

2

The number of blows taken by the executioner to sever the head of Mary, Queen of Scots.

2

The number of dogs hanged for witchcraft during the 1692 Salem witch trials.

2

In Woodbridge, England, in 1799, a judge found two pigs guilty of digging up and eating a corpse. The pigs were sentenced to death by drowning.

3

The number of different types of kisses the Romans had: 'Basium', the kiss on the lips; 'Osculum', a friendly kiss on the cheek; and 'Suavium', a passionate kiss. In fact, the

Romans were so keen on kissing that the Emperor Tiberius was obliged to ban the practice after an epidemic of lip sores.

3

The number of kings Britain had in 1936. As a result of King Edward VIII's abdication, the year 1936 saw three different kings on the throne: his father, George V, himself and his brother George VI. There are two other years when this has happened: 1066 (Edward The Confessor, Harold and William The Conqueror) and 1483 (Edward IV, Edward V and Richard III).

3

The number of baths Louis XIV of France took in his lifetime (and he had to be forced into taking those).

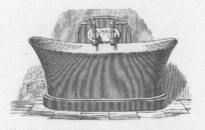

3

The number of Jane Austen's sisters-in-law who died in childbirth.

3

The number of people who were twice awarded the Victoria Cross. Noel Chavasse, Arthur Martin-Leake, both members of The Royal Army Medical Corps, and New Zealander Charles Upham.

3

Robert Todd Lincoln, son of President Abraham Lincoln, was present at the assassinations of three US presidents: Lincoln, Garfield and McKinley.

3

The number of presidents Mexico once had in a single day.

3

The number of times that President Franklin D. Roosevelt was named

Man of the Year by *Time* (the years were 1932, 1934 and 1941, and he is the only person to have been honoured so many times).

4

The number of lions King Henry III kept in the Tower of London.

4

The number of geese Lord Byron took with him everywhere he went.

4

The number of Victoria Cross recipients who have reached the rank of Field Marshal.

5

The number of enemy aircraft a pilot had to shoot down to become an 'ace'.

5

Queen Elizabeth I wasn't buried until five weeks after her death.

6

The number of months Richard The Lionheart spent in England in his lifetime.

6

The number of verifiable deaths in the 1666 Great Fire of London.

7

Napoleon took 14,000 French decrees and simplified them into a unified set of seven laws.

7

The number of years it took Marie Antoinette and King Louis XVI to consummate their marriage.

7

The number of Christian names King Edward VIII had – more than any other king in history. He was baptized with the names Edward Albert Christian George Andrew Patrick David.

7

The number of prisoners in the Bastille when it was stormed in 1789: four forgers, a count committed by his family and two lunatics.

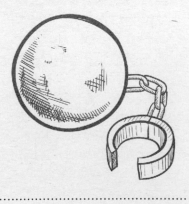

8

In 532 AD, the Byzantine Empire and Sassanid Persia concluded the Eternal Peace treaty. The peace lasted just eight years.

8

Before winning the US Presidential election in 1860, Abraham Lincoln lost eight elections for various offices.

9

The number of days for which Lady Jane Grey reigned as Queen of England in 1554.

9

John William Gott was the last person jailed for blasphemy in Britain (in

1922, he was sentenced to nine months' hard labour after comparing Jesus to a clown).

....

10

According to a study by the International Committee of the Red Cross, the civilian casualty ratio in wars fought since the mid-20th century has been ten civilian deaths for every death of a soldier.

....

10

The number of years in which the Duke of Wellington hadn't seen Kitty Pakenham when he proposed marriage to her in 1806.

....

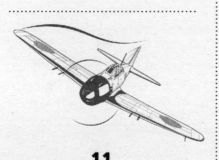

11

The (record) number of missions flown by a Japanese Kamikaze pilot during World War II. The role of the Kamikaze ('Divine Wind') pilots was to fly rudimentary planes packed with explosives straight into Allied ships in what amounted to a suicide mission. However, to complete their task, they had to find an allied ship and, if they couldn't, they returned to base. This is why there was one Kamikaze pilot who flew no fewer than 11 missions and lived through the entire war.

....

11

The number of days lost in 1752 when the calendar changed from the Julian to the Gregorian (Sept 3–13).

....

13

The number of inches (equivalent to 33 centimetres) deemed to be the desirable waist measurement for ladies at the court of Catherine de Medici.

....

13

In the 17th century, the average married British woman gave birth to 13 children.

13

The number of years for which Tchaikovsky was financed by a wealthy widow whom, at her request, he never met.

14

The number of years for which the capital of Portugal was moved to Rio de Janeiro, Brazil, while Portugal was fighting France in the Napoleonic Wars.

14

The number of seconds it took Napoleon's surgeon Baron Dominique Larrey to amputate a leg.

15

In 1705, John Smith was hanged for burglary at Tyburn Tree. After he had been hanging for fifteen minutes, a reprieve arrived and he was cut down. Amazingly, he was revived and managed to recover. As a result of his experience, he became known as John 'Half-Hanged' Smith.

17

All of Queen Anne's 17 children died before she did.

17

The number of times a German military band had to perform the British national anthem on the platform of Rathenau railway station in Brandenburg, on 9 February, 1909. (King Edward VII was struggling inside the train to get into his German Field-Marshal uniform, so the band had to keep on playing the national anthem.)

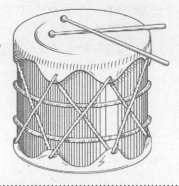

20

The number of minutes for which King Louis XIX ruled France in 1830.

21

The Death Of Twenty-One Cuts was a traditional method of execution in Japan in which the executioner would slice away pieces of the victim's body, killing him with the twenty-first and final cut.

24

The largest number of VCs awarded on a single day (on 16 November 1857, for the relief of Lucknow).

28

The number of different drugs Adolf Hitler was taking daily towards the end of his life.

30

On 29 March 1848, Niagara Falls stopped flowing for 30 hours because of an ice jam blocking the Niagara river.

38

The number of minutes it took Britain to beat Zanzibar in the shortest war in history (in 1896).

39

The number of handkerchiefs used to mop the sweated brow of George IV during his coronation in 1821.

40

The number of governments France had between the two World Wars.

42

The length in kilometres of the Berlin Wall that separated East and West Berlin.

47

The number of MPs arrested on 6 December, 1648, for opposing the trial of Charles I.

47

The number of bleedings Louis XIII's doctors administered to him in the course of just one year (for good measure, they also administered 212 enemas and 215 purgations in the same period of time).

51

The number of founder states of the United Nations in 1945.

55

The number of Turkish ambassadors who refused to remove their hats in the presence of Vlad the Impaler, who consequently had their hats nailed to their heads.

57

The number of countries involved in World War II.

61

The number of centimetres above the ground that the first helicopter reached in 1907, in the first manned helicopter flight.

63

The number of aeroplanes in the Royal Flying Corps at the start of World War I.

63
The Australian soldier
Ringer Edwards survived
63 hours of crucifixion
by the Japanese during
World war II (two other
soliders crucified at the
same time both died).

67
The number of ships that
returned to Spain after
the Spanish Armada – out
of the 130 that set out.

67
The number of nuclear
devices detonated in the
Marshall Islands in the
Pacific Ocean between
1946 and 1958.

70
The number of days
it took to mummify a
dead human being in
ancient Egypt.

82
The percentage of
the workers on the
Panama Canal who
suffered from malaria.

94
The number of years
King Pepi II of Egypt, the
longest ruling king in
history, reigned (from the
age of six until a hundred).

95
The number of 'Theses'
nailed to the door of
the Castle Church in
Wittenberg in 1517 by
Martin Luther (which
were critical of the
Catholic Church).

97

The number of passengers on the Hindenburg airship when it exploded in 1937 (35 of whom died).

116

The number of years for which the 100 Years War lasted. It wasn't one long war like the two World Wars, but a series of wars between England and France. The first one started in 1337 and the last one ended in 1453 – i.e. 116 years. However, there were many brief interludes and two lengthy periods of peace, so the fighting took place, over 81 years.

139

The number of cars produced by the entire American automobile industry during World War II.

150

Marie Antoinette's favourite hairdresser, Leonard, designed a hairstyle called the *loge d'opera* that rose 150 centimetres above the wearer's head.

150

The number of wigs owned by Elizabeth I.

200

The range of a medieval longbow, in metres.

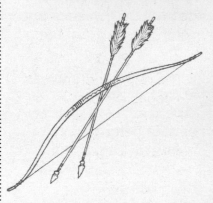

222

The number of offences that were punishable by death in Britain in 1819.

267
The number of words
in Abraham Lincoln's
Gettysburg Address.

269
The (record) number
of minutes that Fidel
Castro's 1960 speech to
the United Nations lasted.

230
It has been calculated that
in the last 3,500 years,
there have been just 230
years of peace throughout
the civilized world.

396
The number of different
rulers Nicaragua had
between 1839 and 1855.

230
The number of miles
Johann Sebastian
Bach once walked to
hear the organist at
Lübeck in Germany.

413
The number of beds
owned by King Louis XIV.

240
The number of wives
inherited by Sultan Murad
IV when he assumed the
throne of Turkey in 1744.
He put each wife into a
sack and tossed them one
by one into the Bosphorus.

634
The number of VCs awarded during World War I (the most for any war).

673
The number of men in the Charge of the Light Brigade.

710
The number of survivors of the *Titanic*.

1,513
The number of passengers and crew lost when the *Titanic* sank.

3,994
The number of oak trees that were used to build Windsor Castle in the 14th century.

4,280
The number of buffaloes killed by Buffalo Bill.

6,000
The number of trees that were used to make HMS *Victory*.

10,500
Sutoku, an Emperor of Japan, copied the Lankauru Sutra, a 10,500-word Buddhist tract, in his own blood.

15,000
The number of dresses Tsarina Elizabeth, who ruled Russia from 1741 to 1762, left in her wardrobe when she died. She never wore the same dress twice.

20,000
The number of Jewish partisans during World War II.

27,500
The number of workers who died during the construction of the Panama Canal.

30,000
The number of werewolf cases that were tried in France between 1520 and 1630.

70,000
To punish a revolt in Persia, Mongol leader Tamerlane had the entire population massacred. He left behind a pyramid of 70,000 skulls piled up outside the city walls.

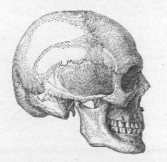

117,000
Abdul Kassem Ismael, the Grand Vizier of Persia in the 10th century, carried his library with him wherever he went. The 117,000 volumes were carried in alphabetical order by 400 camels.

222,570
The number of people who died in the 2010 Haitian earthquake.

830,000
The estimated number of deaths in the 1556 Shensi (China) earthquake.

20,000,000
Fourteen million people were killed in World War I; twenty million died in the flu epidemic that followed it.

SPORTS

0

The number of games Arsenal lost in 2003–4 – thus becoming the first club to win the Premiership without losing a single game.

0

The number of goals Switzerland conceded in the 2006 World Cup Finals while becoming the first team to be eliminated from the World Cup Finals without conceding a single goal.

0

The number of players Tottenham Hotspur had sent off in a Football League match between 27 October 1928 and 4 December 1965.

0

The number of tries scored by Jonah Lomu against South Africa despite facing them 13 times.

1

The number of times animals were killed on purpose in an Olympic event (the 1900 Olympics featured pigeon shooting).

1

The number of prime ministers to own a Derby-winning horse (Earl Rosebery, 1894).

1

The number of England international footballers to wear glasses in an international (James Mitchell 1924).

1

The number of rugby union players to wear a monocle in internationals (Dolway Walkington of Ireland – though he did remove it when he needed to make a tackle).

1

The number of athletes to win an Olympic sprint gold medal while chewing gum (Thomas 'Eddie' Tolan, 200 metres, 1932).

1

The number of jockeys to win a race after death. In 1923, Frank Hayes suffered a fatal heart attack in the middle of a race at Belmont Park in New York. However, his horse, Sweet Kiss, didn't know this and carried on running to win the race with the lifeless jockey still on board.

1

The number of first-class cricketers to have been dismissed off consecutive deliveries in the same first-class match. This happened in 1946 when Glamorgan were playing the touring Indians. Peter Judge was bowled by the last bowl of the Glamorgan innings and the Indians invited the county to follow-on. At this point, the Glamorgan captain, Wilf Wooller, decided to waive the 10-minute interval between innings and instructed the last pair to remain in the middle and open the innings: in other words, he was reversing the entire batting order. The first ball of Glamorgan's second innings saw Peter Judge bowled again (by the very same bowler) – thus becoming the only cricketer to be bowled by consecutive balls in the same match.

1

The number of first-class cricketers to die on the *Titanic* (the American John Thayer).

2

The number of sports in which the team has to move backwards to win (tug of war and rowing – backstroke is *not* a team sport).

2

The number of goals AFC Bournemouth scored against Manchester United – without reply – in 1984.

2

The Czech team that reached the 1934 World Cup Final contained players from just two clubs.

2

Actor Matthew Perry was ranked number two at tennis in Ottawa at the age of 13.

2

The number of Test matches that have been tied – the first between West Indies and Australia at Brisbane in 1960–1 and the second between Australia and India in Madras in 1986–7.

2

The number of dead heats in the Derby. The first dead heat was in 1828 and the two horses concerned raced again later that afternoon. There was a second in 1884 and the two jockeys met in the weighing room and decided to share the prize money.

2

The number of holes-in-one achieved by 75-year-old Peter Wafford in the same round in 2010. The chances of such a feat are estimated at 1 in 67 million.

3

English referee Graham Poll mistakenly handed out three yellow cards to Croatia's Josip Šimunić in a 2006 World Cup match against Australia.

4

In the history of professional boxing, only four men have been knocked out in the first eleven seconds of the first round.

4

The number of balls in a game of croquet: black blue, red and yellow.

4

The number of clubs in the Football League with names starting and ending with the same letter: Liverpool, Charlton Athletic, Northampton Town and Aston Villa.

5

The quickest booking in a Football League/Premiership match was after just five seconds. The culprit was Vinnie Jones playing for Sheffield United against Manchester City. He was booked again later in the match and was therefore sent off.

5

The number of one-eyed men in the 1920 France–Scotland rugby union match. One of them was the remarkable Prop Marcel-Frédéric Lubin-Lebrère who, just a few years earlier in World War I, had lost an eye and also had 23 pieces of shrapnel removed from his body. He later became Mayor of Toulouse.

5

The number of Football League teams for which the footballer Kevin Bremner scored during the 1982–3 season.

5

The number of batsmen who have been left stranded on 99 not out in a Test match. Four of them had either already made test centuries or would go on to do so. However, 99* turned out to be the England cricketer, Alex Tudor's highest test score.

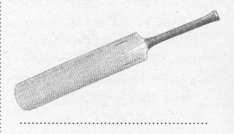

6

The number of past, present or future England captains in the Southampton line-up of the early 1980s: Alan Ball, Mick Channon, Kevin Keegan, Mick Mills, Peter Shilton and Dave Watson.

10

From a complete stop,
a human is capable of
outrunning a Formula One
car for about ten metres.

10

The number of ways
in which a batsman
can get out in cricket:
caught, bowled, leg
before wicket (LBW), run
out, stumped, handling
the ball, obstructing the
field, hit the ball twice, hit
wicket, and timed out. He
can also retire. The first
five are by far the most
common. In Test cricket,
for example, only seven
batsmen have ever been
out for handling the ball,
and only one (Sir Len
Hutton) for obstruction
of the field. No Test
batsman has ever been
dismissed for hitting the
ball twice or been timed
out. A batsman is allowed
to hit the ball a second
time if it's threatening
to hit his stumps but
not at any other time.

10

The number of first-
name initials boasted
by A.R.R.A.P.W.R.R.K.B.
Amunugama who has
more initials than any
other first-class cricketer.

11

A semi-final wrestling
bout in the 1912 Olympics
lasted for 11 hours. The
winner, an Estonian
named Martin Klein, was
so exhausted that he
couldn't take part in the
final contest the next day.

11

James Gordon of Rangers
was selected to play for
the club in all 11 positions,
including goalkeeper,
during his career at Ibrox
Park from 1910 to 1930.

12

The number of clubs that made up the Football League when it was first established in the 1888–9 season: Preston North End, Aston Villa, Wolverhampton Wanderers, Blackburn Rovers, Bolton Wanderers, West Bromwich Albion, Accrington (no relation to Accrington Stanley), Everton, Burnley, Derby County, Notts County and Stoke City. Preston North End won the first title without suffering a single defeat.

13

The number of cricketers who have scored double hundreds in Test matches but finished on the losing side. Twelve of the thirteen batsmen suffered this fate once but it happened to the West Indian cricketer, Brian Lara, on three separate occasions.

13

The number of bowlers who have taken a wicket with their very first ball in Test cricket. Over a hundred bowlers have taken a wicket with their very last ball in Test cricket. No bowler appears on both lists.

13

The number of officials in a tennis match (ten linesmen, one net official, one foot-fault official and an umpire).

14

The number of countries that competed in the first Modern Olympics (in 1896).

14

The number of days the Timeless Test between England and South Africa lasted for in 1939. Remarkably, it wasn't finished: England needed 42 more runs to win, and had five wickets in hand, but the team's boat was due to sail home the next day, and so the game was called off.

18

The number of times that Greek skier Antoin Miliordos fell before crossing the finish line of the downhill event at the 1952 Winter Olympics – backwards.

18

The number of men who died from injuries sustained on the American football field in 1905 – which prompted President Theodore Roosevelt to demand measures to make the game safer.

20

The number of players sent off during a match played in Paraguay between Sportivo Ameliano and General Caballero in 1993.

22

In 1983, the British racing driver John Watson became the only driver to win a Formula 1 Grand Prix from as far back as 22nd on the grid.

25

When a baseball is hit really hard, it momentarily changes shape by as much as 25 per cent.

30

The number of minutes for which the first League match to be played under floodlights (Portsmouth v. Newcastle on 22 February 1956) was held up when the fuses on the floodlights failed.

32.66

The weight in kilograms of Hart Massey, the Oxford cox in the 1939 Boat Race. There's now a 55kg lower weight limit for coxes.

35

The longest tandem was designed for 35 riders. It was twenty metres long and weighed as much as a small car.

36

The longest gap – in years – between appearances in the Olympic Games. Ralph Craig (US 100 metres 1912) competed again in the 1948 Olympics at the age of 59 as an alternate in the US yachting team.

36

The record number of goals scored in a first-class British soccer match – by Arbroath 36 v. Bon Accord 0 on 12 September 1885. Incredibly, on the very same day and in the very same competition – the Scottish Cup – Dundee Harp beat Aberdeen Rovers 35–0. According to the referee in the Dundee Harp v. Aberdeen Rovers game, the final score was 37–0 but the club secretary of Dundee Harp – the winning team, remember – reckoned that it was only 35–0 so the referee went with the lower figure.

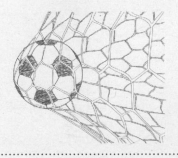

45

The first international Test match took place in 1877. It was between Australia and England. Played in Melbourne, it finished with Australia winning by 45 runs. 100 years later, in 1977, a centenary Test match between the two countries was played in Melbourne. Australia beat England by 45 runs – the precise same margin by which they had won the inaugural match 100 years earlier.

52

When Ted Hough was signed by Southampton FC in 1921, his 'transfer fee' was a round of drinks (which amounted to 52 pints of beer).

59

The number of centimetres above his own height that Franklin Jacobs high-jumped in 1978 – a height differential record that still stands.

66

The number of runners in the 1929 Grand National.

68

The percentage of professional ice-hockey players who have lost at least one tooth.

85

A first-class soccer game has an average of 85 throw-ins – almost one a minute.

87

The number feared by superstitious Australian cricketers because it's 13 short of a century.

99.94

The final Test average of the Australian batsman, Sir Donald Bradman. In his last Test innings (against England at the Oval in 1948), he needed just four runs to be able to retire with a batting average of precisely one hundred. He was bowled second ball by Eric Hollies for a duck. This left him with a test average of 99.94 – still considerably higher than his nearest competitor.

100

Prior to 1900, prize fights lasted up to 100 rounds.

101.2

Maria Sharapova's loudest grunt was measured at 101.2 decibels (louder than a motorcycle or a lawnmower), on 21 June 2005 at Wimbledon Centre Court.

103

The oldest player to 'score his age' in a game of golf was C. Arthur Thompson (1869–1975) of Victoria, British Columbia, Canada, who carded 103 in 1973.

108

The number of stiches on a baseball.

110

A badminton shuttlecock travels at a speed of over 110 mph.

111

The number feared by superstitious English cricketers (known as a Nelson) probably because the figure looks like three stumps without their bails. Very superstitious batsmen will hop on one leg when the score is 111 (222, 333 etc.).

122

The number of successive victories by Ed Moses in the 440-metre hurdles.

138

The world's longest competitive tennis match took place at Wimbledon in 2010, when the American player John Isner beat the Frenchman Nicolas Mahut after playing for eleven hours and five minutes over three days. The reason it went on so long is because in the final set, a player must win by two clear games. The score in their final set was 70–68 – that's 138 games played over 8 hours 11 minutes – making it the longest set in history in both time and games.

147

The maximum break in snooker (15 reds each followed by the black and then all the colours). In fact, with free balls after a foul shot, 155 is technically the highest score possible. The highest break ever recorded was 151 by Cliff Thorburn, who achieved his score with the benefit of a foul shot from his opponent.

156

The highest speed on skis ever recorded was 156 mph by the Italian Simone Origone in April 2006 in Les Arcs, France.

200

The number of spectators who turned up to watch the first Wimbledon championship in 1877 (they paid one shilling each; the first champion, Spencer Gore, won 12 guineas).

223

The (record, for a man) number of matches Jean Borotra played in his Wimbledon career.

250

Eddie Arcaro, one of the greatest jockeys in the history of American horse racing, rode 250 losers before winning his first race.

286

The weight in kilograms of the heaviest sumo wrestler – the Hawaiian, Konishiki Yasokichi – known as the Dump Truck.

311

The number of competitors (all male) at the first modern Olympic Games.

325

In 1930, England's Andy Sandham scored Test cricket's very first triple century – 325 – in what was his very last Test match.

326

The (record) number of matches Martina Navratilova played in her Wimbledon career.

350

The number of football fans who died in a 1964 riot after an equalizer by Peru was disallowed (the referee said afterwards, 'Anyone can make a mistake').

375

The number of full members of the All England Lawn Tennis and Croquet Club, which owns and runs Wimbledon. There are also honorary members, including past singles champions.

1,007

The length in yards of the world's longest golf hole: the par-6, sixth hole at Chocolay Downs Golf Course in Marquette, Michigan.

3,000

The number of cows it takes to supply the US National Football League (NFL) with enough leather for a year's supply of footballs.

46,001

The most pushups ever performed in one day.

52,000

The number of balls used at the Wimbledon tennis tournament each year.

250,000

Keepy-uppy is the art of juggling a football using your feet, knees, legs, chest and head without allowing the ball to hit the ground. The men's record for the longest keepy-uppy was set by Briton Dan Magness who kept a football off the ground for 24 hours at London's Covent Garden in May 2009. No one was counting but it was estimated that he hit the ball 250,000 times. The record he broke had been set in August 2003 by Brazilian Martinho Eduardo Orige who had kept a football in the air for 19 hours and 30 minutes.

20,000,000
The number of golf balls lost in water hazards on British golf courses every year.

GAMES

3

The number of years it took 100 designers to develop the computer game Sims 3.

3

In ten-pin bowling, the feat of achieving three consecutive strikes is called a Turkey.

5

When Richard Pavelle's Rubik's Cube fell into a swimming pool, he dived in after it and solved the puzzle while only coming up for five gulps of air.

7.5

A bowling pin only has to tilt 7.5 degrees to fall down.

17

If done perfectly, any Rubik's Cube combination can be solved in 17 turns.

21

The number of pips on a die.

24

The number of points on a backgammon board.

28
The number of
properties for sale on
a Monopoly board.

30
The number of languages
Scrabble is produced in.

30
If all the Lego in the
world were evenly
distributed, we would
each receive 30 pieces.

42
The number of eyes
in a pack of cards.

51
The world record for
skimming stones is 51 skips.

52
The number of errors
Parker Brothers claimed
Monopoly had when
originally rejecting it.

53
The percentage of
British homes that
have a Scrabble set.

64
The number of squares
on a chess board.

68
The number of inches
off the ground that
the bull's-eye on a
dartboard must be.

78
The number of cards
in a tarot deck.

121
The number of countries in which Scrabble is produced.

126
The highest number of points that you can get in the first go – when there are no other letters on the board – in a game of Scrabble (doing the word SQUEEZY or the word QUARTZY).

126
Since the Slinky first went into production, more than three hundred million have been sold worldwide and in excess of fifty thousand tons of wire have been used. If you joined them all up, they would encircle the Earth more than 126 times.

136
The number of standard tiles in a Mahjong set.

240
The number of white dots in a Pac-Man arcade game.

300
The perfect score in a game of tenpin bowling.

324
The number of potential murder combinations in a game of Cluedo, which has six potential suspects, six possible weapons and nine different places where the murder could have been committed (6 x 6 x 9 = 324.)

666
The sum of all the numbers on a roulette wheel.

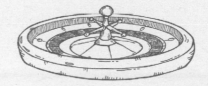

1,680

The number of hours of the longest-recorded Monopoly game (more than 70 days).

2,200

The biggest game of pass-the-parcel was played in Singapore in 1998. It involved thousands of people removing over 2,200 sheets of wrapping paper from a parcel.

26,040

In the game Monopoly, the most money you can lose in one travel around the board (normal game rules, going to jail only once) is £26,040.

2,598,960

The number of possible hands in a five-card poker game.

318,979,564,000

The number of possible combinations of the first four moves in a game of chess.

1,929,770,126, 028,800

The number of different colour combinations possible on a Rubik's Cube.

SCIENCE & NATURE

1
The number of litres of used motor oil that could ruin one million litres of fresh water.

1
If you yelled for 8 years, 7 months and 6 days, you would have produced enough sound energy to heat one cup of coffee.

1.3
A 'scruple' is an apothecary weight equal to twenty grains, or about 1.3 grams.

2.8
The pH of Coca-Cola.

7
The PH value of pure water. More is alkaline and less is acidic.

7.5
The percentage of copper in sterling silver.

10
The percentage of all medicine estimated by the World Health Organization to be fake.

13

The number of
essential vitamins.

13.5

The percentage of
scientists who are women.

18

Sound travels 18 times
faster through steel than
it does through the air.

24

The number of carats
in pure gold.

35

There is enough graphite
in the average pencil to
draw a line 35 miles long.

70

The percentage of all
living organisms in the
world that are bacteria.

71

The weight of the biggest
gold nugget in kilograms –
found in Australia in 1869.

85

The percentage
of gynaecologists
who are men.

90

The percentage of all the
scientists who ever lived
who are alive today.

98

The percentage of times
a computer program
correctly diagnosed
people with various
medical problems. By
contrast, the (real life)
doctors were right in
just 78% of cases.

200

The number of times more gold in the world's oceans than has been mined in our entire history.

250

A scientific satellite needs only 250 watts of power to operate.

322

Airbags explode at 322 kilometres per hour.

450

The number of years it takes for a plastic bottle in the ocean to degrade completely.

760

The weight in nanograms of the average smell.

982

Charcoal is made by grinding up wood and heating it to 982 degrees Celsius, creating a carbon substance known as char.

3,000

When glass breaks, the cracks move at speeds up to 3,000 miles (4,827 km) per hour.

3,422

The number of degrees Celsius at which Tungsten melts (it has the highest melting point of any metal).

4,000

The number of grains of sugar needed to fill a teaspoon.

7,000
The number of cherries on
the average cherry tree.

50,000
The number of years it is
estimated that a plastic
container can resist
decomposition for.

15,000
The number of known
species of orchid (20
per cent of them can
be found in Brazil).

1,250,000
The number of orchid
seeds required to
make up one gram.

15,000
Because of gravity,
it is impossible for a
mountain to be higher
than 15,000 metres.

2,000,000
The number of hydrogen
atoms it would take to
cover the full stop at the
end of this sentence.

200,000
The weight in tons of
the average iceberg.

3,100,000
The number of separate
parts in a Boeing 767.

10,000,000
The number of bacteria
in a litre of milk or in
one gram of soil – or,
indeed, where you rest
your hands on a desk.

70,000,000

The largest flowering plant is 70 million times bigger than the smallest.

9,192,631,770

A second is defined as being 9,192,631,770 times the time it takes for a cesium-133 atom to swap hyperfine levels in its ground state.

METEOROLOGY

-40
The number of the degree at which Celsius is exactly the same temperature as Fahrenheit.

1
The number of people who have been killed by meteorites – (Manfredo Settala in 1680 is the only person in all recorded history to have been killed by one).

2
The temperature drop in degrees Celsius during a total solar eclipse.

3
The number of hours on 14 August 1979 for which a rainbow was visible over North Wales.

9
A full moon is nine times brighter than a half-moon.

10
The number of minutes it takes for a hurricane to release more energy than all the world's nuclear weapons combined.

14
The rate in miles per hour a typical rain-drop falls at.

15
In the UK, a drought is officially defined as

15 consecutive days with less than 0.2 millimetres of rainfall.

20
Very hard rain pours down at the speed of 20 mph.

40
The number of tons of space dust that fall on the Earth every day.

40
The number of days of rain that will follow if it rains on St Swithin's Day (according to superstition).

40
A rainbow can occur only when the sun is 40 degrees or less above the horizon.

57.8
The highest temperature in degrees Celsius ever recorded anywhere in the world – in Al 'Aziziyah, Libya, on 13 September 1922.

60
The number of minutes it can take a snowflake to fall from a cloud to the surface of the Earth.

82
The percentage of people hit by lightning who are male.

92
The number of people who were killed by giant hailstones – each weighing up to one kilogram – in Bangladesh in 1988.

100
The number of times lightning strikes the Earth every second.

1,800
The number of thunderstorms on Earth at any moment in time.

12,000
The square mileage of the largest iceberg ever recorded (it was 200 miles long and 60 miles wide – which is larger than Belgium).

100,000
If you are struck by lightning once, you are 100,000 times more likely to be struck a second time.

500,000,000
The number of litres of rain that can fall in a single thunderstorm.

7,000,000,000
The number of particles of fog it would take to fill a teaspoon.

10,000,000,000
A one-day weather forecast requires about 10 billion mathematical calculations.

ASTRONOMY

2.5
If you were standing on the planet Mercury, the Sun would appear 2.5 times larger than it appears on Earth.

-178
The surface temperature on Saturn in degrees Celsius.

7
The number of lunar and solar eclipses (in total) that are possible in any one year.

-145
The surface temperature on Jupiter in degrees Celsius.

2
At over a thousand times bigger than Earth, Jupiter is more than two times as large as *all* the other planets put together.

8.5
The number of minutes it takes for light to get from the Sun to Earth.

9.8

The number of hours a day on Jupiter (at its equator) lasts.

10

The number of seas on the Moon.

10.56

The miles-per-hour surface speed record on the Moon. It was set in a lunar rover.

11

The number of years it takes for the magnetic poles of the Sun to switch in a process known as 'Solarmax'.

12

The number of people who have set foot on the Moon.

20

The number of seconds of fuel Apollo 11 had left when it landed.

21

Summer and winter on Uranus each last for 21 years.

88

The number of (known) constellations of stars.

142

The number of years it would take to reach the Sun if it were possible to drive through space at 75 mph (120 km/h). At that same speed it would take more than 38 million years to reach the closest stars.

152
The number of minutes
spent by Neil Armstrong
on the Moon.

225
The average wind
speed on the planet
Jupiter in mph.

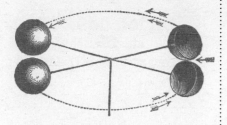

248.4
The number of years
it takes for Pluto to
orbit the Sun.

462
The surface
temperature on Venus
in degrees Celsius.

530
The extra fuel in kilograms
needed at lift-off for
every additional kilogram
carried on a space flight.

1,200
The average wind speed
on the planet Saturn.

3,537
The number of years
it would take to
walk to the Sun.

25,000
The number of light years
we are from the Canis
Minor Dwarf Galaxy – our
nearest galactic neighbour.

1,300,000
The Sun is 1,300,000 times
bigger than the Earth
in volume.

1,598,000
The number of miles
travelled each day
by the Earth.

29,000,000
A car travelling at 100 mph would take more than 29 million years to reach the nearest star.

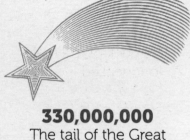

330,000,000
The tail of the Great Comet of 1843 was 330 million kilometres long. (It will return in 2356.)

360,000,000
Every year the sun loses 360 million tons.

5,000,000,000
The Milky Way galaxy contains five billion stars larger than our sun.

10,000,000,000
Most stars shine for at least ten billion years.

CREEPY-CRAWLIES

1

The number of foods that cockroaches won't eat (cucumbers).

1

The number of insects with retractable antennae (the snail).

1.8

The distance in miles the male gypsy moth can 'smell' the virgin female gypsy moth.

2

The number of years tarantulas can go without eating.

2

The neck of the male longnecked weevil is two times the length of its body.

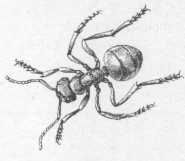

2

The number of weeks ants can survive underwater.

3

The number of body parts all insects have – a head, a thorax, and an abdomen. They also have six (jointed) legs, and two antennae that they use as sensors.

3

The number of years
for which a snail can
sleep without eating.

4

The average spider will
spin more than four miles
of silk in a lifetime.

4

Slugs have four noses.

4

Fleas can suck blood
continuously for four
hours. However, their
stomachs can't hold
all that blood so they
excrete what they can't
hold and their offspring
gobble up the leftovers.

5

The number of hearts
an earthworm has.

8

Relative to its size, the
ordinary house spider is
eight times faster than
an Olympic sprinter.

9

The number of hours
it takes for an ant
to walk a mile.

10

The number of times
by which the world's
termites outweigh the
world's humans.

10

The percentage of the
world's food supply
consumed by insects.

11

The number of brains the
silkworm has (although it
uses only five of them).

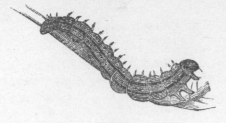

18
The number of hours
that cockroaches carry
on releasing methane
gas after they die.

12
The number of eyes
a caterpillar has.

24
Adult earwigs can float in
water for up to 24 hours.

12
Snails mate only once in
their lifetime, but it can
take up to 12 hours.

27
The temperature in
degrees Celsius to which
a butterfly warms up
its body before flying.

12
The number of hours
it can take a spider
to eat a large fly.

15
The number of minutes
a cockroach can
survive underwater.

17
The cicada, a fly found
in Africa, spends 17 years
of its life sleeping. In the
two weeks it's awake, it
mates and then dies.

30
The ant can pull thirty
times its own weight.

32
The number of
brains a leech has.

33

The longest recorded
tapeworm found in
a human body was
33 metres long.

45

The total weight in
grams of all the honey
made by a single bee
in its entire lifetime.

47

The number of teeth
a mosquito has.

50

The ant can lift 50
times its own weight (it
can also pull 30 times
its own weight).

50

Some cockroaches are
so fast they can run 50
times their own body
length per second.

50

Queen termites can live
for up to 50 years under
the right conditions.

55

In 1864, a bootlace worm
measuring 55 metres was
washed up on the shore
in Scotland. Even at their
usual length – 30 metres
– bootlace worms are
incredibly long, although
they're rarely thicker
than one centimetre.

61

The speed in centimetres
per hour of a snail.

75
The number of flowers visited by a honey bee during a collection trip.

80
The percentage of creatures on Earth that have six legs.

99
The percentage of baby tarantulas killed by their own mothers.

100
The weight in grams of the heaviest insect – the goliath beetle.

150
Gram for gram, a bumblebee is 150 times stronger than an elephant.

200
The sensors on the feet of a red admiral butterfly are 200 times more sensitive to sugar than the human tongue.

200
Scorpions can withstand 200 times more nuclear radiation than humans can.

350

The flea can jump 350 times its body length.

572

The number of wing flaps per second made by a mosquito.

650

The number of different types of leeches (though only the Hirudo medicinalis is used for medical treatments).

750

The approximate number of legs a millipede has.

1,800

There are over 1,800 known species of flea.

2,000

The number of eggs a queen bee lays a day.

2,000

The number of muscles a caterpillar has in its body (we humans have 656).

2,500

The North American black-and-orange Monarch Butterfly is the only insect known to be capable of flying over 2,500 miles. It flies between continents in its migration.

3,000

Different types of lice.

11,400

The number of times a minute a bee flaps its wings.

18,000
Different species
of butterfly.

25,000
The number of
teeth a snail has.

30,000
From hatching to
pupation, a caterpillar
increases its body size
more than 30,000 times.

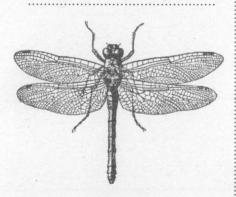

30,000
The number of lenses in
each of a dragonfly's eyes.

80,000
A large swarm of locusts
can eat 80,000 tons
of corn in a day.

350,000
The number of known
species of beetle –
with millions more
waiting for names.

1,000,000
There are one million ants
for every person in the
world.

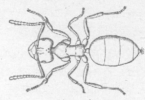

2,000,000
Honey bees must gather
nectar from two million
flowers to make one
pound of honey.

6,000,000
The average person
has a one in six million
chance of being killed
by a bee sting.

8,000,000
There are up to eight
million worms in the soil
of each hectare of forest.

100,000,000

Ants evolved from wasps more than 100 million years ago.

150,000,000

The number of young a female mosquito can produce in one year.

200,000,000

For every person on Earth, there are an estimated 200 million insects.

700,000,000

Blood-sucking hookworms inhabit 700 million people worldwide.

6,000,000,000

The number of dust mites in a typical bed.

SEA CREATURES

0.01
The slowest sea creature is the seahorse, which moves along at about 0.01 mph.

1
Dolphins sleep with one eye open.

1
Typically only one per cent of the 500 million eggs laid annually by the American Oyster reaches maturity.

0.014
A sperm whale has a brain that weighs 7.8 kg (17.2 lb), but that is only 0.014% of its body weight.

1
The number of weeks it takes a shark to grow a new set of teeth.

1
The number of bites or squirts from the Blue-Ringed Octopus needed to cause immediate paralysis and death in minutes.

2
The number of tons of food a large whale eats daily.

2
Female whales can live up to two times longer than male whales.

2
Sea otters have two coats of fur.

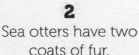

2.5
The giant squid can weigh up to 2.5 tons.

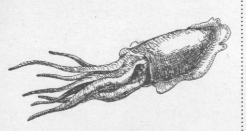

3
Galapagos turtles can take up to three weeks to digest a meal.

3
Dolphins have three stomachs.

3
A whale can swim for three months without eating.

3
The number of hearts an octopus has.

4
The number of minutes it takes to die after being stung by the 'Sea Wasp' or Box Jellyfish.

4
Atlantic salmon are able to leap 4 metres high.

5

The electric eel's power is so great that it can shock its victims at a distance of five metres.

5

The number of piranha fish it would take to consume a horse and rider in seven minutes.

6

The number of milliseconds that it takes frogfish to suck prey into their mouths – which is too fast for other animals to see.

7

The Weddell seal can travel underwater for seven miles without surfacing for air.

7.5

The length, in metres, of a baby blue whale at birth.

9

Tuna swim at a steady rate of nine mph until they die and they never stop moving.

9

A blue whale's heart beats just nine times per minute.

10

About ten times more men than women are attacked by sharks.

10

The number of tentacles a squid has.

14

Dolphins have the best sense of hearing in the natural world: they can hear 14 times better than humans and can pick up an underwater sound from several miles away.

35
The largest species of seahorse measures 35 centimetres.

40
The percentage of seahorses – which are genetically monogamous within a breeding period – that get 'divorced' and 'remarried' between reproductive periods.

45
The number of years lived by Goldie, the oldest known goldfish who was won at a fairground in 1960.

50
The bluefin tuna can swim at 50 mph.

54
The weight in kilograms

of a swordfish found – with its sword still intact – inside a mako shark weighing 330 kilograms.

60
Manatees live for up to 60 years.

85
The percentage of all life on Earth that is plankton.

90
The percentage of all living things that live in the sea.

90
The European freshwater mussel lives for at least 90 years.

90

Seals sleep for just 90
seconds at a time.

90

The percentage of all fish
caught that are caught in
the northern hemisphere.

95

The percentage of a
jellyfish that is water.

100

The number of eyes
a scallop has – and
they're all blue.

100

The male sea lion can
have more than 100 wives.

140

The number of years that
George the Lobster was
estimated to have lived.

188

The number of decibels
produced by the whistle
of the blue whale (the
loudest sound produced
by any creature).

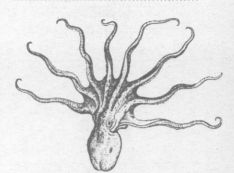

200

Octopus arms
typically have about
200 suckers each.

200

The number of metres
for which the flying fish
can stay airborne as
it leaps into the air to
escape its predators.

400

The number of volts an
electric eel produces.

800
Fish that live more than 800 metres below the ocean surface don't have eyes.

2,000
A baby grey whale drinks enough milk to fill more than 2,000 bottles a day.

3,000
To reach the rivers and lakes where they spend most of their lives, many newborn eels swim non-stop for up to 3,000 miles.

12,000
The number of miles that grey whales migrate each year – further than any other mammal.

21,000
There are about 21,000 varieties of fish on Earth.

27,000
The number of taste buds the catfish has (more than any other creature).

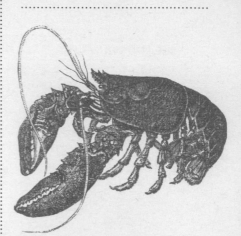

150,000
The number of eggs a lobster can lay at one time.

224,000
The blue whale weighs approximately as much as 224,000 copies of *Moby Dick*.

1,000,000
A 15-year-old tuna travels one million miles in its lifetime.

2,000,000

The female starfish produces two million eggs a year, of which 99 per cent are eaten by other fish.

5,000,000

The average cod deposits five million eggs in a single spawning.

20,000,000

The number of seahorses that are harvested each year for folk medicinal purposes. As a consequence, the world seahorse population has dropped 70% in the past 15 years.

170,000,000

The number of neurons in the brain of an octopus. A human brain has more than 100 *billion*.

BIRDS

1

The number of birds that provide us with leather (the ostrich).

1

The number of eggs the female condor lays every two years.

1

The number of birds with nostrils at the end of their bills (the kiwi).

1

The number of birds – or, indeed any species other than human beings – that French kiss. The bird in question is the white-fronted parrot. However, after they touch tongues, the male vomits on the female's chest.

1

The number of birds that can fly backwards (the hummingbird).

1

The number of birds that can walk upright (the penguin).

2

The number of weeks for which Goldie the golden eagle escaped from London Zoo in 1965. He became a national celebrity during his liberty when he flew around the roads and trees of Regent's Park.

2

The number of toes an ostrich has on each foot.

2

The number of minutes by which a flock of homing pigeons beat the Liberal MP Sir William Edge in a race between London and Leicestershire. He lost when his train was delayed.

2

The number of times a year a penguin has sex.

3

Pigeons process visual information three times quicker than we do.

3

The number of eyelids a duck has.

3

The number of minutes for which grebes can stay underwater.

5
The waste produced by one chicken in its lifetime can supply enough electricity to run a 100-watt bulb for five hours.

10
The number of miles a condor can fly without flapping its wings.

4
The number of eyelids an ostrich has. The inner lids are for blinking and keeping the eyeballs moist, the outer lids for attracting potential mates.

4.6
The wingspan in metres of the world's largest pelican.

5
The broad-tailed hummingbird drinks five times its own body weight in nectar every day.

10
The number of bird species – of the 10,000 bird species so far discovered – that have been domesticated.

10
The number of times a woodchuck breathes during hibernation.

11

The number of kilograms a father Emperor penguin loses while withstanding the Antarctic cold for 60 days or more to protect his eggs, which he keeps on his feet, covered with a feathered flap. During this entire time, he doesn't eat a thing.

11.4

The pouch under a pelican's bill holds up to 11.4 kg of fish and water.

12

The male European house wren will build up to 12 nests to entice a female.

12

The percentage of an egg's weight accounted for by its shell.

13

The longest-recorded flight of a chicken is 13 seconds.

20

The number of times a second a woodpecker can peck.

20

Parrots usually have a vocabulary of up to 20 words (though there have been parrots with vocabularies of over one hundred words).

20
The number of
minutes for which the
Emperor penguin can
hold its breath.

24
The number of chicken
eggs you'd need to
equal the weight of a
single ostrich egg.

25
The number of years
for which the world's
oldest duck lived.

28
The number of
orders of birds.

34
The weight in kilograms
of the largest turkey
ever produced.

40
The number of minutes
it takes to soft-boil
an ostrich egg.

40
The number of different
muscles in a bird's wing.

42
The number of days
it takes for an ostrich
egg to hatch.

50
An owl can see a mouse
moving from fifty metres
away by a light no brighter
than candlelight.

50
The number of years flamingoes can live up to.

55
The speed in mph at which wild turkeys can fly for short distances. Domesticated turkeys (farm-raised) cannot fly. Wild turkeys are also fast on the ground, running at speeds of up to 30 miles per hour.

70
The number of years an Andean condor can live up to.

70
The number of different (common) breeds of chicken.

75
The percentage of wild birds that die before they reach six months old.

90
The percentage of bird species that are monogamous.

156
The adult male ostrich, the world's largest living bird, can weigh up to 156 kg.

300
The number of different calls used by the crow.

600

Pigeons can fly up to 600 miles in one day.

2,000

The ostrich egg is 2,000 times bigger than the smallest egg, which is the hummingbird's. An ostrich egg weighs 1.2 kilograms. A hummingbird egg weighs half a gram.

24,843

One Laysan albatross, tracked by biologists at Wake Forest University, flew more than 24,843 miles in flights across the North Pacific to find food for its chick in just 90 days — a flight distance equivalent to circling the globe.

REPTILES

1

The number of lungs most snakes use for breathing.

2

The number of heads the spiny thorny devil lizard has. However, one of these is a fake. If it's seriously threatened and it's a matter of life or death, it will extend its fake head to its attacker while hiding its real head between its front legs.

3

The number of eyes a lizard has. The third eye is thought to help them differentiate between dawn and dusk as it is sensitive to blue and green lights.

2

The number of years for which female alligators – unlike other reptiles – protect their young after hatching.

3

The number of types of snakes on the island of Tasmania (all of which are deadly poisonous).

6

The Komodo dragon – the largest lizard in the world – has such a keen sense of smell that it can locate a dead or dying animal from up to six miles away.

11

The speed in miles per hour a crocodile can run at.

12

The number of months for which some snakes can live without eating.

12

The number of hours that sex between snakes can last.

13

Milk snakes lay about 13 eggs – in piles of animal manure.

26

The chameleon releases its tongue at 26 body lengths per second – faster than the human eye can see. It hits its prey in about 0.03 seconds.

28

The number of minutes an iguana can stay underwater.

118
The number of sheep
the tiger snake can
kill with its venom.

3,000
The number of teeth
an alligator can go
through in a lifetime.

200,000
The inland taipan is the
world's most poisonous
snake. The venom it
produces in one bite
would be enough to kill
200,000 mice.

ANIMALS

0
The number of times giraffes kneel (i.e. they never do).

0.0000000000002
You can tell if a skunk is about if you smell only 0.0000000000002 grams of its spray.

0.15
The percentage of the African elephant's weight taken up by its brain (that weighs 7.5 kilograms).

1
Hamsters blink one eye at a time.

1
Cheetahs make a chirping sound that can be heard one mile away.

1
A newborn kangaroo weighs less than one gram and is small enough to fit in a teaspoon.

1
The number of animals – apart from man – that can catch leprosy (the armadillo).

1

The number of animals
– apart from man –
that can stand on their
heads (the elephant).

1

The number of kicks
it would take a giraffe
to kill a lion.

1

The number of breeds of
bear with blue tongues
(the Black Bear).

1

Ferrets have just one
type of blood group –
unlike other mammals,
which have several
different types.

1

The number of cats
that can't retract their
claws (the cheetah).

1.82

When a giraffe is born,
it has to fall some 1.82
metres to the ground.

2

The number of weeks for
which sheep can survive
buried in snow drifts.

2

The number of years
for which a baby beaver
stays with its parents.

2

The antler of a male
moose measures about
two metres across.

2

The number of mammals
that lay eggs (the
platypus and the echidna:

213

the mothers nurse their babies through pores in their skin).

2

The number of hours, on average, elephants sleep each day.

2

The number of years for which the African elephant is pregnant.

2

The number of weeks for which the hamster is pregnant.

2

The number of dogs that survived the sinking of the *Titanic*.

2

Lions can't roar until they reach the age of two.

3

The number of eyelids a camel has (necessary to protect themselves from blowing sand).

3

The number of days for which rats can tread water.

3

The greater dwarf lemur – as found in Madagascar – always gives birth to three babies (i.e. triplets).

3

The number of strides it takes greyhounds to reach their top speed of 45 miles per hour.

3

In India, the term 'man-eating' is only applied to tigers that have killed three or more people. Man-eating tigers are usually those that are too old to capture wild animals.

3

The baby caribou can outrun its mother when it's just three days old.

3.8

The thickness in centimetres of a hippopotamus's skin – almost bulletproof.

4

The number of knees an elephant has (the only animal thus endowed).

4

The placement of a donkey's eyes in its head enables it to see all four feet at one time.

4

The number of babies an armadillo has at a time (always all of the same sex).

4

A dairy cow produces four times its weight in manure each year.

4

The number of cat species that can roar (and they don't purr): lions, leopards, tigers and jaguars.

4

The number of metres
a jackrabbit can travel
in one bound.

5

Skunks can withstand five
times the snake venom
that would kill a rabbit.

5

The woolly mammoth,
extinct since the Ice
Age, had tusks almost
five metres high.

5

An adult lion's roar is
so loud, it can be heard
up to five miles away.

6

The number of
days a sloth takes to
digest its food.

6

The night vision of
tigers is six times better
than that of humans.

6

The distance in miles a
cow can smell odours.

6

The number of distinctive
sounds horses use
to communicate.

7

A horse eats about
seven times its own
weight a year.

7.5

The number of miles
a pig can cover when
running at top speed.

8

The distance in metres an Australian red kangaroo can jump in one bound.

9

The number of litres of water an elephant's trunk can hold.

10

The duration in seconds of the average bout of intercourse between chimpanzees.

10

There are ten times more horses than people in Mongolia.

12

All the pet hamsters in the world are descended from one female wild golden hamster found with a litter of 12 young in Syria in 1930.

13

Some dinosaurs had tails up to 13 metres.

18

The distance in feet pumas can leap.

18

The number of naps a rabbit takes a day.

18.5

The average number of hours sleep armadillos get a day.

19
The number of kittens in the largest cat litter ever recorded (although four were still-born).

20
The number of miles from which a polar bear can smell a human being.

20
The number of hours a day a lion sleeps.

20
The number of minutes a day a giraffe sleeps.

20
A lion in the wild usually makes no more than 20 kills a year.

20
The number of years a squirrel could live for in captivity. However, their life span in the wild is only about one year as they fall prey to disease, predators, malnutrition, cars and humans.

22
The number of hours a day a koala sleeps.

23
The amount of dung an average elephant produces a day in kilograms.

24
A farmer introduced 24 wild rabbits into Australia in 1859. There are now

an estimated 300 million rabbits there.

25
The percentage of all the horses in the US that died in an epidemic in 1872.

26
Domestic cats purr at about 26 cycles per second, the same frequency as an idling diesel engine.

28
The number of years on average naked mole rats live – longer than any other rodent; seven times longer than mice of the same size.

29
The oldest dog that ever lived was 29 years old.

30
Both elephants and chimps have around 30 different sorts of sounds to communicate different things to one another.

32
The number of muscles a cat has in each ear.

32
The common little brown bat of North America is, for its size, the world's longest lived mammal with life-spans of up to 32 years.

34
The oldest age a cat has ever reached.

38
The length in centimetres of a bread knife swallowed by a 45-centimetre

Collie/Staffordshire terrier crossbreed dog named Kylie. She swallowed the knife with the sharp end in her stomach and the blunt end sticking out of her mouth and lived to tell the tale.

45
The number of minutes for which beavers can hold their breath.

38
The maximum number of kilograms of bamboo a day eaten by a giant panda.

45
The fastest animal on four legs is the cheetah, which can accelerate to 45 miles per hour in two seconds.

45
The weight in kilograms of an elephant's ear.

40
You can tell the sex of a horse by its teeth. Most males have 40, females have 36.

46
The highest in inches a rabbit has ever jumped.

42
The number of teeth a bear has.

47
The speed in miles per hour rabbits have been known to reach.

50
The length in centimetres of a giraffe's tongue, which it uses to clean its ears.

50
The number of sheep faces a sheep can remember.

50
The number of hours it takes for a snake to digest a frog.

61
The maximum speed (mph) a pronghorn antelope can run at.

62
The oldest horse in the world – Billy, a barge horse – lived to be 62 years. Horses generally live on average between 20 and 25 years.

67
The number of African elephants that would equal the weight of a Boeing 747.

70
The percentage of an elephant that is water.

75
The length of tunnel in metres a mole can dig in a single night.

75

The percentage of its own weight that a bat can eat in a single evening.

80

The percentage of their vocalizations that hippopotamuses do underwater.

90

The percentage of the hunting that the female – rather than the male – lion does.

90

The approximate weight in kilograms of a giraffe when born.

95

The percentage of the world's laboratory mice that are descended from mice born in the Jackson Laboratory in Bar Harbor, Maine.

98

The percentage of DNA that chimpanzees and humans have in common.

99.9

The percentage of all the animal species that have ever lived on Earth that were extinct before the coming of man.

100

The number of vocal sounds cats have in excess of (dogs only have about ten).

100

The number of babies one mouse can give birth to in a year.

100

Approximately, the amount of blood in gallons an elephant has.

146

The number of
endangered mammals in
Indonesia – more than any
other country (followed by
India, China and Brazil).

200

The amount of food in
kilos an African elephant
can eat per day. Their
diet consists of twigs,
leaves, grasses and fruit.

150

There are more than
150 breeds of horses in
the world. With some
11 million horses within
its borders, China has
more horses than
any other country.

200

The number of trees
a beaver can chop down
in a year.

205

The number of
bones in a horse.

160

The number of times per
minute anteaters can
stick out their tongues.

225

The number of
bones in a mouse.

190

The rate at which a
hedgehog's heart beats
per minute. This drops
to 20 beats per minute
during hibernation.

258

The range of genetic
diseases cats can
suffer from.

260

The breaths per second a dormouse breathes before going into its winter hibernation.

273

The number of hedgehogs killed on British roads on an average day. This may explain why hedgehog numbers have declined by about a third in the past decade.

300

The number of people injured in an average year in half a million car accidents involving animals – killing 30,000 of them in the process – most of them are deer.

345

The number of squirts it takes on average to yield a gallon of milk from a cow's udder.

500

The amount in litres of methane gas a cow produces a day.

500

The individual hair of a chinchilla is so fine that 500 of them equal the thickness of a single human hair.

500

The ferret was domesticated more than 500 years before the cat.

500

The approximate number of spines on a hedgehog.

600

The number of bugs per hour a bat eats through the night every night.

650

The number of times a minute the heart of a mouse beats.

1,000

Koko, a gorilla born at San Francisco Zoo in 1971, mastered up to 1,000 words in sign language.

1,000

There are fewer than 1,000 Bactrian camels left in the wild.

1,700

Represents rodents as the largest order of mammals. Bats are second with about 950 species.

2,080

The number of warren exits found in a single colony housing 407 rabbits – sociable creatures often found living in large groups in underground burrows.

4,000

The number of years since the last new animal was domesticated.

7,000

By the age of six months, a baby pig will have increased its birth weight 7,000 times.

10,000

The number of insects a single toad can eat in the course of a summer.

15,000
Two rats can become the progenitors of 15,000 rats in less than a year.

30,000
The approximate number of ants the South American giant anteater eats per day.

30,000
The approximate number of quills on the average porcupine.

35,000
The estimated number of stray dogs in Moscow (some of which live in its underground stations).

40,000
The number of muscles in an elephant's trunk (there are no bones).

110,000
The number of venom extractions from the coral snake it would take to fill a one-litre container.

200,000
The number of glasses of milk a cow gives in her lifetime.

400,000
The number of farmed foxes in Finland – which is the world's leading producer of fox pelts.

5,000,000
The number of years for which the first dinosaur – the Staurikosaurus – survived.

6,000,000
The number of cats in the UK. The most popular breeds in the UK are Persian long hair, Siamese and British short hair.

16,000,000
The number of animals that assisted the armed forces during World War I.

34,000,000
The number of kangaroos in Australia – some twice as many as the human population.

150,000,000
The number of years dinosaurs lived on Earth (that's about 75 times longer than humans have lived on Earth).

WORDS

0

The number of words starting with the letter 'x' in Samuel Johnson's *Dictionary of the English Language*, published in 1755.

0

The number of words in the English language that rhyme with silver or orange.

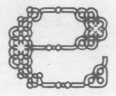

0

The number of times the letter 'e' appeared in the 1939 novel *Gadsby*. Written by Vincent Wright who tied down the 'e' on his typewriter but otherwise didn't cheat, this is an example of a lipogram. It ran to 50,000 words without using any 'e's at all and it was a good book by all accounts. Sadly, Wright, who took 165 days to write it – simply to prove that it could be done – died on the day of the book's publication. Note also, the French author Michel Thaler who published a 233-page novel without any verbs.

0
The number of
Hawaiian words that end
in consonants.

1
The number of capital
letters in the alphabet
that have exactly
one endpoint (P).

1
The number of words
in the English language
that when capitalized
are changed from nouns
or verbs to a nationality
(polish/Polish).

1
The number of
common words with
five consecutive
vowels (queueing).

1
The number of
common four-letter
words consisting
entirely of nonrepeating
consecutive letters (rust).

1
The number of letters
required to turn the one-
syllable word 'are' into a
three-syllable word (area).

1
The number of UK cities
with letters that are all
in the first half of the
alphabet (Lichfield).

1
The number of letters
required to turn the one-
syllable word 'smile' into
a three-syllable word
(simile).

229

1

The number of
common words with
four 'g's (giggling).

1

The number of letters in
the alphabet with more
than one syllable (w).

1

The number of words with
four pairs of double letters
in a row (subbookkeeper).

2

The number of letters in
the shortest sentence in
the English language (Go!).

2

The number of six-letter
words that begin and end
with a vowel and have
no other vowels between
(asthma and isthmi).

2

The number of countries
with three alphabetically
consecutive letters in
their names (Afghanistan
and Tuvalu).

2

The Chinese ideogram
for 'trouble' shows
two women living
under one roof.

2

The number of times
each vowel appears in the
word ultrarevolutionaries.

3

The number of words
that end in the letters
'cion' (coercion, scion,
and suspicion).

3

The number of (relatively) common words that have a letter that repeats six times: degenerescence (six e's), indivisibility (six i's), and non-announcement (six n's).

3

The number of six-letter words with letters in alphabetical order without repetition (abhors, almost and biopsy).

4

The number of words that end in 'dous' (tremendous, horrendous, stupendous and hazardous).

4

The English language has four times as many words as the French.

5

The number of vowels in Mozambique.

5

The words 'abstemious' and 'facetious' contain all five vowels in alphabetical order.

5

The words 'uncomplimetary', 'subcontinental' and 'duoliteral' contain all five vowels in reverse alphabetical order.

5

The number of syllables in the word 'monosyllable'.

6

The number of words in the English language with the letter combination 'uu': Muumuu, vacuum, continuum, duumvirate, duumvir and residuum.

Knightsbridge

6

The number of consonants in a row in the words 'borschts', 'latchstring' and 'weltschmerz'. Note also, Knightsbridge, which has the most consonants in a row.

7

The number of letters in the word 'wronged' – the longest word with its letters in reverse alphabetical order without repetition.

8

The number of letters in the word 'couscous' – the longest word in the English language in which you can't tell visually if it's been written in all upper case or all lower case letters.

8

The number of letters in the word 'feedback' – the shortest word that contains the letters A, B, C, D, E, and F.

8

The number of letters in the word 'cabbaged' – the longest word with letters that can all be 'played' on a musical instrument (using the notes A, B, C, D, E and G).

9

The number of letters in the word 'spoonfeed' – the longest word with its letters in reverse alphabetical order.

9

The number of letters in the word 'startling' – the longest word from which you can keep removing a letter to form new words along the way: startling, starting, staring, string, sting, sing, sin, in, I.

9

The number of words that make up a quarter of all words used in English: the, of, and, to, it, you, be, have, will.

10

The number of letters in the word 'skepticism' (the US spelling of the word 'scepticism') – the longest word that requires you to alternate hands when typing.

12

The number of letters in the Hawaiian alphabet.

therein

13

The number of words spelled with consecutive letters contained in the word 'therein': the, he, her, er, here, I, there, ere, rein, re, in, therein, herein.

13

The percentage of letters in any given book that are 'e'.

15

The number of letters in the word 'fifteen-lettered' making it autological in that it truly describes itself (see also 'unhyphenated' which is autological in that it is indeed unhyphenated).

15

The only 15-letter word that does not repeat a letter is 'uncopyrightable'.

15

The number of letters in the name of the tennis player, Goran Ivanešević – the longest name of a celebrity that alternates consonants and vowels.

18

The number of letters in the word 'strengthlessnesses' – the longest word in the English language with just one (repeated) vowel.

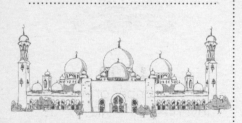

18

The number of letters in United Arab Emirates – the longest name of a country consisting of alternating vowels and consonants.

18

The number of letters in the words 'conversationalists' and 'conservationalists',

which, between them, make up the longest pair of anagrams in the English language.

18

The number of letters in the word 'overnumerousnesses' – the longest English word which consists of only letters that lack ascenders, descenders, and dots in lower case.

32

The number of letters in the Polish alphabet (including three variations of the letter 'z').

36

The number of letters in the Albanian alphabet (but it has no 'w').

36

The number of letters in the word 'hippopoto-monstrosesquippedaliophobia', which means the fear of long words.

40

The number of letters George Bernard Shaw believed we should have in our alphabet (and he left money in his will to help bring this about).

72

The number of letters in the Cambodian alphabet.

400

The words-per-minute rate at which American tobacco auctioneers can speak.

812

The number of three-letter words in the English language.

1,000

The number of words that make up 90 per cent of all writing.

1,700

The number of words invented by Shakespeare.

5,700

The number of characters on an old-fashioned Chinese typewriter.

95,940

The number of words in the longest will in history – which was drawn up for Frederica Cook, an American woman. When it was proved at London's Somerset House in 1925, it consisted of four bound volumes. Incredibly, she only had about £25,000 to leave.

7,100,553
Putting this number into a calculator and then turning it upside down produces the words ESSO OIL.

71,077,345
Putting this number into a calculator and then turning it upside down produces the words SHELL OIL.

378,193,771
Putting this number into a calculator and then turning it upside down produces the word ILLEGIBLE.

53,177,187,714
Putting this number into a calculator and then turning it upside down produces the word HILLBILLIES.

531,607,017,818
Putting this number into a calculator and then turning it upside down produces the word BIBLIOLOGIES.

531,607,018,036
Putting this number into a calculator and then turning it upside down produces the word GEOBIOLOGIES.

531,607,055,076
Putting this number into a calculator and then turning it upside down produces the word GLOSSOLOGIES.

LOVE & MARRIAGE

(in fact, it was less than one day: it was just after the wedding ceremony).

1

The number of days the marriage of Zsa Zsa Gabor and Felipe De Alba lasted.

0

The number of marriages entered into by Chopin, Casanova, Gershwin, Sartre, Descartes, Kant, Nietzsche and Larkin.
In other words, they couldn't muster up a single marriage between them.

1

The number of days the marriage of Eva Bartok and William Wordsworth lasted

1

The number of months the marriage of Burt Lancaster and June Ernst lasted.

3

The number of months the marriage of Nicolas Cage and Lisa Marie Presley lasted.

4

The number of months the marriage of Joanna Lumley and Jeremy Lloyd lasted.

4

The number of days the marriage of Ethel Merman and Ernest Borgnine lasted.

6

The number of days the marriage of John Heard and Margot Kidder lasted.

6

The number of times Sir Rex Harrison married.

6

The number of times Gloria Swanson married.

7

The number of times Lana Turner married.

7

The number of times Richard Pryor married (twice to two of his wives).

7

The number of times Stan Laurel married (three times to the same woman).

7

The number of times Claude Rains married.

8

The number of times Dame Elizabeth Taylor married – twice to the same man: Richard Burton.

8

The number of times Mickey Rooney married.

8

The number of times Alan Jay Lerner married.

8

The number of times Artie Shaw married.

9

The number of times Victor Hugo – previously a virgin – had sex on his wedding night.

9

The number of times Pancho Villa married.

9

The number of days the marriage of Cher and Gregg Allman lasted.

9

The number of months the marriage of Marilyn Monroe and Joe DiMaggio lasted.

9

The number of times Zsa Zsa Gabor married.

11

The number of months the marriage of Helen Hunt and Hank Azaria lasted.

11
The number of
months the marriage
of Paul Simon and
Carrie Fisher lasted.

17
The number of years
younger Raymond
Chandler was than his
wife, Pearl Bowen.

17
The number of years
younger Clark Gable
was than his first wife,
Josephine Dillon.

32
The number of years older
Joan Collins is than her
husband, Percy Gibson.

56
The number of times
Richard and Carol Roble
re-married each other.

67
The number of years
the world's longest
engagement lasted for.
Octavio Guillen and
Adriana Martinez were 15
when they got engaged.

104
The number of women
that the bigamist Giovanni
Vigliotto married in
14 different countries
between 1949 and 1981.

9,000
The number of wives of
King Mongut of Siam. He
was quoted as saying he
only loved the first 700.

10,000
The number of women
that Maigret author
Georges Simenon claimed
to have slept with.

PURE
TRIVIA

1

The number of museums dedicated to strawberries (it's in Belgium).

1

The number of foods that astronauts do not have to treat and dehydrate before eating when flying in space (pecans).

1

The number of museums dedicated to penises (it's in Iceland).

1

The number of inanimate symbols in the zodiac (Libra, the Scales).

2

The number of times a Punch and Judy performer has to have swallowed his swazzle before he can consider himself a professor.

2

The number of people in the US Baseball Hall of Fame who had nothing

to do with baseball
– i.e they didn't play,
coach, own a team etc.
(Abbott and Costello).

4

The number of
sandwiches Charles
Lindbergh took with
him on his famous
transatlantic flight.

3

The number of years
in jail to which Juan
Bernaus was sentenced in
Argentina for switching the
'Ladies' and 'Gents' signs
round on public toilets.

4

Alice Johnson, a 23-year-
old American waitress,
won a car in Santa Fe
after kissing it for 32 hours
and twenty minutes
in a 1994 competition.
She loosened four
teeth in the process.

3

The number of times the
Pepsi-Cola Company
was declared bankrupt
in its infancy.

3

The number of teeth Stalin
had left when he died.

4

In 1997, Robert Brett, a
Californian who wasn't
allowed to smoke at
home, left his entire
fortune to his wife
provided that she smoked
four cigars a day for the
rest of her life.

6

The number of men shaved in one minute by champion barber Robert Hardie in 1909.

5

Disneyworld is bigger than the world's five smallest countries.

6

The number of months an outbreak of contagious laughter in Tanganyika lasted for in 1962 (it caused schools to be closed).

5

The number of magic beans in *Jack and the Beanstalk*.

5

The number of children Ralph and Carolyn Cummins had between 1952 and 1966 – all of whom were born on 20 February.

5

The number of years it took Marva Drew of Iowa, US, to type all the numbers from one to one million.

6

The number of people who were taken to hospital after a rocket fired from a bonfire organized by the Orpington Liberal Club landed on people attending the Orpington Conservative Club's bonfire.

6

Ten-gallon hats only hold about six pints of water.

6

The percentage of Beatles autographs in circulation that are reckoned to be genuine.

7

The number of people who have gone over Niagara Falls in a barrel and survived.

8

In 1992, Frank Perkins of Los Angeles made an attempt on the world flagpole-sitting record. Suffering from flu he came down eight hours short of the 400-day record. Just to make things even worse, he then discovered that his sponsor had gone bust, his girlfriend had left him and his phone and electricity had been cut off.

9

The number of hexagonal pencils that can be made with the same wood it takes to make eight round pencils.

9.17

The greatest distance in metres anyone has spat a dead cricket from their mouth.

10

The percentage of all European babies estimated to have been conceived on Ikea beds.

10
The number of hours Albert Einstein slept every night.

11
In the 1994 World Cup, Bulgaria fielded a team in which all 11 players' last names ended with the letters 'ov'.

11
The number of points on Kermit the Frog's neck collar.

11
The lowest whole number not mentioned in a single Beatles song.

12
In the 1820s, people believed that travelling by trains at a speed of more than 12 mph would cause mental problems.

12
The number of centimetres the giant water lily grows every day.

14.5
In 1994, Mars took out full page ads in newspapers announcing their 'New Biggest Ever Mars'. The 'Emperor'-sized Mars bar was 14.5 kg of 'thick chocolate, glucose and milk' but it was only 'on sale' for one day: 1 April.

18
The number of bicycles eaten by the Frenchman Michel Lotito (he also ate 15 supermarket trolleys, six chandeliers, two beds and a pair of skis).

20

The percentage of Nobel Prize-winners who are/were Jewish or of Jewish ethnicity.

22

The number of children fathered by Siamese Twins Chang and Eng Bunker.

23

The percentage of photocopier faults caused by people photocopying their buttocks.

30

The length in metres of the world's longest limousine (it has a built-in swimming pool and costs more than $5,000 a day to rent).

36

The toll paid in cents, in 1928, when Richard Halliburton swam the Panama Canal.

37

Surgeons who spent at least three hours a week playing computer games made 37% fewer mistakes in surgery than surgeons who didn't.

44

The number of ways – according to Swiss scientists – to build an Ikea wardrobe without the instructions (though only eight of these ways will result in a correct and safe construction).

50

The percentage of all lingerie purchases returned to the store.

50
The percentage of the world's population that has seen at least one James Bond movie.

53
The number of hours Marge Simpson was in labour with Bart.

54
The length in miles of Fifteenmile Creek.

62
The percentage of all email that is spam.

65
The percentage of Elvis impersonators who are of Asian descent.

78
The record number of kisses Adrian Chiles received in one minute (air kisses didn't count).

80
The estimated percentage of time capsules that will be lost before they get the chance to be opened.

85
While in Alcatraz, Al Capone was inmate #85.

97
The percentage of Canadians who said that they would not borrow a toothbrush if they forgot to pack their own.

97
The weight in pounds (equivalent to 44 kg) of the weakling in the Charles Atlas ads who got sand kicked in his face.

107

The number of seconds it took a team of three people to open 300 beer bottles.

112

The length of a metal coat hanger, in centimetres, when straightened.

150

The number of calories used per hour by banging your head against a wall.

174

The number of cats eaten in a year by Charles Domery.

178

The number of legs in the song 'The Twelve Days of Christmas' (1 partridge, 2 doves, 3 hens, 4 colly birds, 6 geese, 7 swans, 8 maids with a cow each, 9 ladies, 10 lords, 11 pipers and 12 drummers).

225

The number of different ways the Dutch town of Leeuwarden can be spelled.

254

The number of expired American Express cards that went into the making of the American Express Gold-card dress worn at the 67th Academy Awards by costume designer Lizzy Gardiner.

600

The number of different ways to make love according to the Marquis de Sade.

603

The number of years before the creation of Diet Coke that a girl born in West Riding, Yorkshire, was named Diot Coke.

Coke was a corruption of 'Cook', and 'Diot' was a diminutive of Dionisia.

950
The number of times a sixty-eight-year-old South Korean woman took her written driver's test before passing it in November 2009.

1,313
Donald Duck lives at 1313 Webfoot Walk, Duckburg, Calisota.

2,354
Two of the Basque Separatist ETA members convicted of the 1987 Zaragoza Barracks bombing were each sentenced to 2,354 years in prison.

5,000
The number of crocodile skins Australia exports annually from its crocodile farms.

7431
Homer Simpson's PIN number.

11,967
The record for the largest number of people involved in a line dance – held in Singapore.

25,000
Juan Potomachi, an Argentinian, left the equivalent of £25,000 to his local theatre on condition that they used his skull when performing *Hamlet*.

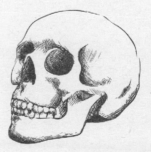

41,000

The number of ping-pong balls the singer Harry Richman insisted on carrying on a plane for fear that it would crash into the sea.

60,000

In 2007, during a month-long performance-art piece in Chicago, the Iraqi-born artist Wafaa Bilal had more than 60,000 paintballs fired at him (remotely by people pressing a button on their computers).

100,000

Professional gambler Brian Zembic had size-38C breast implants inserted into his chest to win a $100,000 bet.

137596

Sam Spade's licence number.

181,000

Philip Grundy, a dentist, left his dental nurse £181,000 on condition that she didn't wear any make-up or jewellery or go out with men for five years.

250,000

The estimated number of cigars smoked by Sir Winston Churchill in his lifetime.

800,000
The number of fan
letters Mickey Mouse
got in the year 1933.

1,500,000
The number of people
who logged on to the
website of the cheddar
cheese Wedginald to
watch it mature.

3,085,209,600,000
The number of rolls of
wallpaper it would take to
cover the Sahara desert.

INDEX

(some indexed items appear more than once on a given page)